国家级一流本科专业建设成果教材

Hydraulics
水力学
（英汉双语）
第二版
English-Chinese

吴慧芳　主编

肖智兴　蒋海艳　陆梦恬　副主编

化学工业出版社

·北京·

内容简介

作为双语教材，本书英语与汉语并重，每一章节先用英文阐述，后面紧跟中文内容。全书共分为9章，主要内容包括流体的性质（Properties of Fluids）、流体静力学（Fluid Statics）、流体运动及其基本方程（Fluid Flow and Basic Equations）、量纲分析和相似原理（Dimensional Analysis and Similarity Principle）、流动阻力和能量损失（Flow Resistance and Energy Loss）、有压管流（Pressure Flow）、明渠流（Flow in Open Channels）、渗流（Seepage Flow）。

本书可作为理工科院校给排水科学与工程、环境工程、土木工程等专业的双语教材，也可以供相关专业的教师、研究生、科学技术工作者及工程技术人员参考。

图书在版编目（CIP）数据

水力学 ＝ Hydraulics：英、汉 / 吴慧芳主编；肖智兴，蒋海艳，陆梦恬副主编. -- 2版. -- 北京：化学工业出版社，2025.2.--（国家级一流本科专业建设成果教材）. -- ISBN 978-7-122-46836-9

Ⅰ．TV13

中国国家版本馆CIP数据核字第2024RE9026号

责任编辑：满悦芝　　　　文字编辑：杨振美
责任校对：张茜越　　　　装帧设计：张　辉

出版发行：化学工业出版社
　　　　　（北京市东城区青年湖南街13号　邮政编码100011）
印　　装：北京建宏印刷有限公司
787mm×1092mm　1/16　印张13¼　字数332千字
2025年2月北京第2版第1次印刷

购书咨询：010-64518888　　　售后服务：010-64518899
网　　址：http://www.cip.com.cn
凡购买本书，如有缺损质量问题，本社销售中心负责调换。

定　　价：49.80元　　　　　　　　　版权所有　违者必究

前 言

双语教学是把英语作为第二语言的教学，其内涵是培养学生用英语阅读、思考、学习专业知识的能力，关键是通过英语学习专业知识，并在此过程中提高发现问题、分析问题、解决问题的能力。教育部出台的《关于加强高等学校本科教学工作提高教学质量的若干意见》，明确要求高等院校在本科教育方面积极推动使用英语等外语进行公共课和专业课教学。教学改革是教学工作永恒的话题，现代教学的一个主要目的就是要紧跟学科发展的步伐，培养有发展前途的创新型人才。具备良好的外语（主要是英语）运用能力，是创新型人才走向世界的一个重要条件。高校开展双语教学，就是在保留本学科特点和保证教学质量的前提下，尽量多地营造双语的学习氛围，最大限度地给学生们提供运用外语的空间和时间，激发学生对学习外语知识的兴趣，提高英语的运用能力和学科知识的掌握水准，实现学科交融、互动、双赢，最终促进学生们综合素质、学识水平的提高。双语教学的目标是要追求人才素质的长期效应，并不停留在学习外语专业术语或词汇上。从长远看，双语教学就是要培养学生的外语应用能力，提高终身学习、获取信息的能力，使学生们在本科毕业后能够自如地继续学习专业知识，了解国内外本学科以及相近学科领域的前沿与最新进展。它要求教学安排从外语语言点向专业知识点逐步转变，外语教学从传统的以教师授课、灌输知识为主的方式向与专业教学相适应的注重研究方法、提出问题的方式过渡，使学生能够通过自我探究获得心理上的外语氛围。所以，普通高校实施双语教学对学生的成才十分重要，意义深远。如何开展双语教学已经成为我们努力探索和尝试的重要课题。

编者在开展水力学双语教学实践过程中，感受到教材的选择与使用存在一些困难。英文原版教材的内容与课程教学大纲要求的内容有较大的差异。有些知识点在英文教材里没有，英文教材中的部分例题和习题的单位采用的是英制（BG），而不是国际单位制（SI），在使用的过程中带来了一些障碍。英文教材中有部分内容是我国中学授课内容，比如气体状态方程这部分内容在高中物理课里就学习了。因此，很有必要编写一本适合我国大学生知识背景的双语教学教材。本教材的编写得到了南京工业大学教材建设项目及给排水专业国家一流专业建设点、江苏省品牌专业建设项目的资助，教材在双语教学实践过程中形成的讲义基础上，吸收国内外有关教材的优点编著而成。

本书的编写在确保符合科学性的前提下，力求严格、准确、形象、清晰，把知识点提炼出来，对理想流体与实际流体、牛顿流体与非牛顿流体、动力黏性系数与运动黏性系数等概念单独列出进行阐述。

全书由吴慧芳任主编，肖智兴、蒋海艳、陆梦恬担任副主编，邓风、郑超凡参加编写。全书共分为9章，具体分工如下：南京工业大学吴慧芳编写第1、3、4、6章；郑超凡编写第2章；邓风编写第5章；蒋海艳编写第7章；陆梦恬编写第8章；肖智兴编写第9章。全书由吴慧芳统稿。

本书的出版还要感谢化学工业出版社的大力支持，感谢诸多师生的支持。

由于编者水平有限，书中不妥之处在所难免，敬请读者批评指正。编者邮箱 wuhuifang@163.com。

编者
2024 年 10 月 于南京

目 录

■ **Chapter 1　Introduction** ·· 1
第 1 章　绪论

1.1　Tasks of Fluid Mechanics ·· 1
1.1　流体力学的任务 ·· 1

1.2　History of Fluid Mechanics ·· 2
1.2　流体力学的发展史 ·· 3

1.3　Research Methods of Fluid Mechanics ································· 4
1.3　流体力学的研究方法 ·· 4

1.4　How to Study Fluid Mechanics Well ···································· 4
1.4　如何学好流体力学 ·· 5

■ **Chapter 2　Properties of Fluids** ·· 6
第 2 章　流体的性质

2.1　Differences between Solids and Fluids ································· 6
2.1　固体和流体的区别 ·· 6

2.2　Continuous Medium Hypothesis ·· 6
2.2　连续介质的假设 ·· 7

2.3　Density and Specific Weight ··· 7
2.3　密度和重度 ·· 7

2.4　Viscosity ··· 8
2.4　黏性 ·· 9

2.5　Compressibility and Expansivity ······································· 14
2.5　压缩性和膨胀性 ·· 14

2.6　Surface Tension ··· 18
2.6　表面张力 ·· 18

2.7　Vapor Pressure ··· 19
2.7　蒸气压 ·· 19

2.8　Forces on a Fluid ··· 19
2.8　作用在流体上的力 ·· 19

Exercises 习题 ·· 21

Chapter 3　Fluid Statics
第 3 章　流体静力学 23

3.1　Static Pressure and Its Characters 23
3.1　静压强及其特性 24

3.2　Differential Equation of Fluid Equilibrium 26
3.2　流体平衡的微分方程式 27

3.3　Basic Equation of Fluid Statics 31
3.3　流体静力学基本方程 32

3.4　Measurement of Static Pressure 35
3.4　流体静压强的测量 37

3.5　Relative Equilibrium of Liquid 38
3.5　液体的相对平衡 38

3.6　Hydrostatic Forces on Plane Surfaces 44
3.6　作用在平面上的静水总压力 44

3.7　Hydrostatic Forces on Curved Surfaces 50
3.7　作用在曲面上的静水总压力 50

Exercises 习题 54

Chapter 4　Fluid Flow and Basic Equations
第 4 章　流体运动及其基本方程 58

4.1　Lagrange's Method and Euler's Method 58
4.1　拉格朗日法和欧拉法 58

4.2　Basic Concepts of Fluid Flow 62
4.2　流体运动的基本概念 63

4.3　Types of Fluid Flow 64
4.3　流体运动的类型 64

4.4　System and Control Volume 66
4.4　系统与控制体 67

4.5　Continuity Equation 67
4.5　连续性方程 68

4.6　Motion Differential Equation of Ideal Fluid 70
4.6　理想流体的运动微分方程 71

4.7　Bernoulli's Equation and Its Application 72
4.7　伯努利方程及其应用 73

4.8　Momentum Equation and Its Application 89
4.8　动量方程及其应用 91

Exercises 习题 95

Chapter 5 Dimensional Analysis and Similarity Principle
第 5 章 量纲分析和相似原理 ··· 99

5.1 Dimensions and Units	99
5.1 量纲和单位	100
5.2 Dimensional Homogeneity	101
5.2 量纲和谐原理	101
5.3 Rayleigh's Method and Buckingham's π Theorem	101
5.3 瑞利法和白金汉 π 定理	102
5.4 Similarity Principle	108
5.4 相似原理	108
5.5 Similarity Criterion	110
5.5 相似准则	111
5.6 Model Experiment	114
5.6 模型试验	115
Exercises 习题	115

Chapter 6 Flow Resistance and Energy Loss
第 6 章 流动阻力和能量损失 ··· 118

6.1 Laminar and Turbulent Flows	118
6.1 层流和紊流	118
6.2 Basic Equation of Uniform Flow	122
6.2 均匀流基本方程式	123
6.3 Laminar Flow in a Round Pipe	125
6.3 圆管中的层流	126
6.4 Turbulent Flow in a Round Pipe	129
6.4 圆管中的紊流	130
6.5 Friction Resistance in Pipeline	137
6.5 管路中的沿程阻力	137
6.6 Minor Resistance in Pipeline	142
6.6 管路中的局部阻力	142
Exercises 习题	147

Chapter 7 Pressure Flow
第 7 章 有压管流 ··· 150

7.1 Single Pipeline	150
7.1 简单管道	151
7.2 Multiple-Pipe Systems	157

7.2 复杂管道 ······ 157
7.3 Application of Pipelines ······ 158
7.3 管道的应用 ······ 160
Exercises 习题 ······ 162

Chapter 8　Flow in Open Channels
第 8 章　明渠流 ······ 164

8.1 Open Channel ······ 164
8.1 明渠 ······ 165

8.2 Calculation Equation of Uniform Flow in Open Channel ······ 165
8.2 明渠均匀流的计算公式 ······ 166

8.3 Most Efficient Cross-section and Allowable Velocity ······ 170
8.3 明渠水力最优断面和允许流速 ······ 171

8.4 Specific Energy and Critical Depth at a Section ······ 176
8.4 断面单位能量与临界水深 ······ 176

Exercises 习题 ······ 180

Chapter 9　Seepage Flow
第 9 章　渗流 ······ 182

9.1 Introduction to Seepage Flow ······ 182
9.1 渗流简介 ······ 182

9.2 Basic Law of Seepage Flow ······ 183
9.2 渗流基本定律 ······ 183

9.3 Steady Uniform and Steady Gradually-Varying Seepage Flows of Groundwater ······ 184
9.3 地下水的恒定均匀渗流和恒定渐变渗流 ······ 184

9.4 Basic Differential Equation and Seepage Lines of Gradually-Varying Seepage Flow ······ 186
9.4 渐变渗流基本微分方程和浸润曲线 ······ 186

9.5 Catchment Passage and Well ······ 191
9.5 集水廊道和井 ······ 191

Comparison between English and Chinese Terms
英汉术语对照 ······ 197

References
参考文献 ······ 203

Chapter 1 Introduction
第1章 绪 论

1.1 Tasks of Fluid Mechanics

As an important branch of modern mechanics, fluid mechanics is the science of fluid balance, mechanical movement laws and the interaction between fluid and the objects around it. It mainly ascertains the distributions of velocity and pressure, energy loss and the interaction force and moment between fluid and solid.

The most studied fluids in fluid mechanics are water and air. Besides, fluids also include vapor as work medium in turbine, lube, water which mixing with mud and sand, blood, melting metal and the gas with complicated components coming from burning, and plasm under high temperature etc. Hydraulics is the physical science and technology of the static and dynamic behaviors of fluids.

Fluid mechanics is an ancient subject with quite wide research field. There are many questions about fluid mechanics in these departments, such as aviation, spaceflight, navigation, astronomy and meteorology, physical geography, water conservancy and hydropower, heat energy and refrigeration, civil engineering and environmental protection, petroleum and chemical engineering, air and liquid transportation, combustion and explosion, metallurgy and mining, biology and oceanography, war industry and nuclear energy, machine manufacture, mechanical engineering, etc.

1.1 流体力学的任务

流体力学是近代力学的一大分支，它是研究流体的平衡和机械运动规律以及流体与周围物体之间相互作用的科学，主要是确定流体的速度分布、压强分布与能量损失，以及流体与固体相互间的作用力与作用力矩。

流体力学中研究得最多的流体是水和空气。除水和空气以外，流体还包括作为汽轮机工作介质的水蒸气、润滑油、含泥沙的水体、血液、熔化状态下的金属、燃烧后产生的成分复杂的气体、高温条件下的等离子体等。水力学是研究液体静态和动态行为的一门物理科学和技术。

流体力学是一门历史悠久的学科，拥有极为广阔的研究领域。例如：航空、航天、航海、天文气象、自然地理、水利水电、热能制冷、土建环保、石油化工、气液输送、燃烧爆炸、冶金采矿、生物海洋、军工核能、机械制造、机械工程等部门都有许多流体力学问题。

1.2 History of Fluid Mechanics

The first contribution to the formation of fluid mechanics was made by Archimedes of ancient Greece. He founded the liquid balance theory including the physical principles of buoyancy and buoyancy stability, and then the basis of the hydrostatics was established.

The stages of formation and development of fluid mechanics can be divided into five periods.

The first period: before the 20th century B. C. Some questions of the projectile motion in fluids were put forward.

The second period: from the 20th century B. C. to the late 17th century A. D.

Archimedes (287-212 B. C.)—quantified theory of buoyancy;

Maliaut—the balance to measure the resistance of kinetic object;

Pasica—the basic formula of hydrostatics;

Leonardo da Vinci (1452-1519)—equation of conservation of mass in one-dimensional steady flow and the turbulence phenomenon.

The third period: from the late 17th century A. D. to the 20th century A. D.

Newton (1642-1727)—laws of viscosity of Newtonian fluid;

Bernoulli (1700-1782)—Bernoulli's theorem for steady flow of incompressible fluids (1738);

Euler—describing method of fluids motion and equations set for motion of inviscid fluids (1775);

Lagrange—stream function;

Reynolds (1842-1912)—Reynolds experiment and Reynolds equation;

Navier (1785-1836)-Stokes (1819-1905)—differential equations for motion of viscous fluids;

D'Alembert—D'Alembert paradox.

The fourth period: from the early to middle period in the 20th century A. D.

Prandtl (1875-1953)—boundary layer theory (1904), to be the single most important tool in modern flow analysis, the father of modern fluid mechanics;

Taylor (1886-1975)—laid foundation for the present state of the art in fluid mechanics.

The fifth period: after the middle 20th century A. D.

Now some questions in fields such as meteorology, oceanography, petroleum, chemical engineering, energy sources, environmental protection and architecture has been researched by using fluid mechanics, and many new branches of fluid mechanics were formed because the fluid mechanics was infiltrated with relative con-

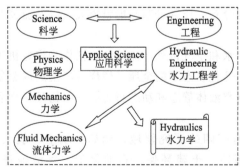

Figure 1-1 The Formation of Hydraulics
图 1-1 水力学的形成

tiguous subjects each other.

Hydraulics is an applied fluid mechanics with hydraulic engineering background, focusing on the theory of fluid flow and its applications in hydraulic engineering. The formation of hydraulics and modern hydraulics can be shown in Figure 1-1 and Figure 1-2.

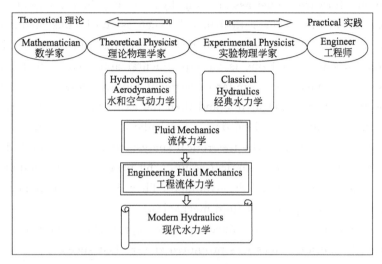

Figure 1-2　The Formation of Modern Hydraulics
图 1-2　现代水力学的形成

1.2　流体力学的发展史

古希腊的阿基米德对流体力学学科的形成做出了重要贡献。他建立了包括浮力的物理原理和浮力稳定性在内的液体平衡理论，奠定了流体静力学的基础。

流体力学的形成和发展大致分为五个阶段。

第一阶段：公元前 20 世纪前，流体中抛体运动问题的提出。

第二阶段：公元前 20 世纪至公元 17 世纪下叶。阿基米德（前 287—前 212）——浮力的定量理论；马里奥特——测量运动物体阻力的天平；帕斯卡——流体静力学的基本关系式；达·芬奇（1452—1519）——一维恒定流质量守恒方程和湍流现象。

第三阶段：17 世纪下叶至 20 世纪初叶。牛顿（1642—1727）——牛顿流体的黏性定律；伯努利（1700—1782）——定常不可压缩流体的伯努利定理（1738 年）；欧拉——流体运动的描述方法和无黏性流体运动的方程组（1775）；拉格朗日——流函数；雷诺（1842—1912）——雷诺实验、雷诺方程式；纳维（1785—1836）-斯托克斯（1819—1903）——黏性流体运动微分方程；达朗贝尔——达朗贝尔佯谬。

第四阶段：20 世纪初叶至中叶。普朗特（1875—1953）——边界层理论（1904），该理论成为现代流体运动简便而重要的分析工具，被称为现代流体力学之父；泰勒（1886—1975）——为流体力学技术发展奠定基础。

第五阶段：20 世纪中叶以后。流体力学开始研究气象、海洋、石油、化工、能源、环保和建筑等领域中的问题，并与有关邻近学科相互渗透，形成许多新分支。

水力学是以水利工程为背景的一门应用流体力学，它研究的主要是流体流动的理论及其在水利工程中的应用。图 1-1 和图 1-2 分别表示了水力学与现代水力学的形成过程。

1.3 Research Methods of Fluid Mechanics

There are three approaches to study fluid mechanics. The first one is a theoretical approach, which puts forward proper assumptions by analyzing the primary and secondary factor of the question, abstracts theoretical model (such as continuous medium, ideal fluid, incompressible fluid etc), and uses mathematical tools to find out the general answer about the fluid movement. The second one is an experimental approach, which summarizes the flow questions in experiment as a similar experiment model, observes the phenomena and determines the data during experiment, and then the experiment results are presumed according to a certain method. The third one is a calculating approach, which draws out the experimental scheme according to the theoretical analysis and experimental observation, inputs the data and calculates the numerical solution by program with computer. All the approaches have some advantages and disadvantages. They need to complement each other in order to promote the development of fluid mechanics.

Modern fluid mechanics has combined the three approaches together organically, and the development of fluid mechanics was promoted at very fast speed.

1.3 流体力学的研究方法

流体力学有三种研究方法。一种是理论方法，分析问题的主次因素之后提出适当的假设，抽象出理论模型（如连续介质、理想流体、不可压缩流体等），使用数学工具寻求流体运动的普遍解。一种是实验方法，将实验流动问题概括为相似的实验模型，在实验中观察现象、测定数据，进而按照一定的方法推测实验结果。还有一种是计算方法，根据理论分析与实验观测拟订实验方案，通过编制程序输入数据，用计算机算出数值解。这三种研究方法各有所长，也各有所短，需要相辅相成才有利于推进流体力学的发展。

现代流体力学已将这三种方法有机地结合起来，流体力学因此得到了飞速发展。

1.4 How to Study Fluid Mechanics Well

(1) Grasping the study method from common to especial   Begin with the general laws for mechanical movement of object, grasp the common type of the basic equation sets, and then make a concrete analysis of concrete problems. The derived equations are only the simplified application of the basic equations under different conditions.

(2) Making great efforts to grasp basic knowledge   Master the basic principles, basic concepts, and basic methods. Do exercises repeatedly and understand them profoundly.

(3) Listening to the teacher carefully and marking the note properly   Listen carefully to the teacher's explanations and treatment methods for the key and difficult points that you think you heed. Record the thought and method of analyzing the questions about the typical example problems in class.

(4) Making good preparations for attending the lecture   Make preparations for difficult chapters, and listen to the teacher's explanations mostly when it is difficult to understand

them by yourself. Moreover, you should read the book with your most energy, but it is the most important that you should understand it, not remember it by rote.

(5) Standardization of solving problems and enhancing basic skills training　Insist on solving questions with physical method. Define the type of the questions and conclude carefully, recognize the knowledge points in it, and then master the thought and method to solve the questions so that you can comprehend by analogy to solve questions and draw inferences about other cases from one instance.

(6) Attaching importance to experiments and doing it yourself　Do experiments stated in the teaching syllabus carefully by yourself, and enforce the understandings, applications and sublimations of the knowledge you have learned. Develop the capability of independent work and lay the foundation for research on scientific experiments in the future.

The physical properties of fluids are determined by the internal reasons of fluid equilibrium and motion regulations, so we should know the concepts and the mainly physical properties of fluids first before the regulations of fluid mechanics are discussed.

1.4　如何学好流体力学

（1）掌握从一般到特殊的学习方法　从物体机械运动的普遍规律出发，掌握一般形式的基本方程组，再根据具体条件分析具体问题。派生方程只是基本方程在不同条件下的简化应用。

（2）在掌握"三基"上下功夫　掌握基本原理、基本概念、基本方法，反复训练，深刻理解。

（3）认真听课，适当记笔记　对自己认为的重点、难点，认真听老师的讲解和处理方法，对典型的课堂例题，应记录分析问题的思路、解题步骤。

（4）做好预习，有准备地听课　对较难的章节，一定要预习，看不懂的地方重点听老师讲解，要把大部分的精力放在看书上，重要的是理解，不要死记硬背。

（5）解题规范化，加强基本功训练　坚持用物理方法解题，认真归纳，明确其中的知识点，掌握解题的思路和方法步骤，达到解题触类旁通、举一反三的目的。

（6）重视实验，亲手去做　对教学大纲中规定的实验都亲自认真去做，加强对所学知识的理解、应用和升华。培养独立的动手能力，也为将来进行科学实验研究奠定基础。

流体的物理性质取决于流体平衡和运动规律的内部原因，因此在讨论流体力学的规律之前，应首先了解流体的概念和流体的主要物理性质。

Chapter 2　Properties of Fluids
第 2 章　流体的性质

2.1 Differences between Solids and Fluids

The molecules of a solid are usually closer together than those of a fluid. The attractive forces between the molecules of a solid are so large that a solid tends to retain its shape. While the attractive forces between the molecules of a fluid are smaller and the intermolecular cohesive forces are not great enough to hold the various elements of fluid together. Hence a fluid will flow under the action of the slightest shear stress and flow will continue as long as the shear stress is present. A fluid may be either a gas or a liquid. The molecules of a gas are much farther apart than those of a liquid.

The appearances caused by these microcosmic differences are as following. The solid has definite volume and shape; the liquid has definite volume but has indefinite shape; the gas has neither definite volume nor definite shape. These are the remarkable differences between solids and fluids.

2.1 固体和流体的区别

固体的分子间距通常比流体的分子间距要小，固体分子间的引力大，因而能够保持一定形状。相比之下，流体间的引力较小，分子间的力不足以使流体保持原状。因此，流体在微小的剪切力作用下就会流动，只要剪切力存在，流动就能持续进行下去。流体可以是气体，也可以是液体，气体分子间的距离比液体要大得多。

这些微观差异导致的宏观表象如下：固体有一定的体积和一定的形状；液体有一定的体积而无一定的形状；气体既无一定的体积也无一定的形状。这就是固体与流体的显著区别。

2.2 Continuous Medium Hypothesis

In dealing with fluid-flow relations on a mathematical or analytical basis, it is necessary to consider that the actual molecular structure is replaced by a hypothetical continuous medium. The number of molecules involved is immense, and the distance between them is normally negligible by comparison with the distances involved in the practical situation being studied. This is called continuum.

In a fluid, a physical body, whose macroscopical scale is very small but microcosmic scale is big enough, has its volume and mass. This is called fluid parcel. Fluid particle is the smallest unit of a fluid whose linear scale can be ignored.

Euler put forward the hypothesis of continuous medium mechanical model in 1753. Without regard to the space of molecules, a fluid is considered to be composed of parcels which have no space between each other and distribute continuously throughout the region in which the liquid lies. When a fluid flows continuously, all the physical variables to represent liquid properties, such as density, speed, pressure, shear stress and temperature and so on are single-valued and continuous differentiable functions on time and spatial variables. Consequently, the scalar and vector fields of all kinds of physical variables come into being. In this way, we can use continuous function and field theory to study the problems about fluid motion and balance successfully.

2.2 连续介质的假设

在处理以数学或分析为基础的流体流动问题时，用假想的连续介质来代替实际的分子结构是很有必要的。流体内部的分子数量非常大，与实际研究的情况相比，它们之间的距离通常是可以忽略的，这就是连续介质的概念。

流体中宏观尺寸非常小而微观尺寸又足够大的任意一个物理实体，具有自己的体积和质量，这就是流体微团。流体质点是流体中可以忽略线性尺度的最小单元。

欧拉在1753年提出了连续介质力学模型的假设。不考虑分子间隙，认为流体由相互间没有间隙的微团组成，连续分布于流体所占据的整个空间；表征流体属性的诸物理量，如密度、速度、压强、切应力、温度等在流体连续流动时是时间与空间坐标变量的单值、连续可微函数，从而形成各种物理量的标量场和矢量场。这样，我们就可以顺利地运用连续函数和场论等数学工具研究流体运动和平衡问题。

2.3 Density and Specific Weight

2.3.1 Density

Density ρ is defined as the mass of the substance per unit volume. The density at a point in fluids is determined by considering the mass Δm of a very small volume ΔV surrounding the point. The density at a point is the limiting value as ΔV tends to zero, that is

$$\rho = \lim_{\Delta V \to 0} \frac{\Delta m}{\Delta V} = \frac{dm}{dV} \tag{2-1}$$

The unit of ρ is kilogram per cubic meter (kg/m^3).

For water at the standard pressure ($1.013 \times 10^5 N/m^2$) and 4℃, the density is 1000 kg/m^3.

2.3 密度和重度

2.3.1 密度

单位体积内物质的质量称为密度。流体内某一点的密度可以认为是该点附近非常小的体积 ΔV 内的质量 Δm，某点的密度就是 ΔV 趋近0时的值。

$$\rho = \lim_{\Delta V \to 0} \frac{\Delta m}{\Delta V} = \frac{dm}{dV} \tag{2-1}$$

密度 ρ 的单位是千克每立方米（kg/m^3）。

在一个标准大气压（$1.013 \times 10^5 \text{N/m}^2$）下，温度为 4℃时，水的密度是 1000kg/m^3。

2.3.2 Specific Weight

The specific weight of a fluid is defined as its weight per unit volume and is denoted as γ.

For homogeneous fluid, the specific weight γ is the ratio of its weight G to its volume V, hence

$$\gamma = \frac{G}{V} \tag{2-2}$$

For heterogeneous fluid, to define specific weight γ at a point, the weight ΔG of fluid in a very small volume ΔV surrounding the point is divided by ΔV and the limit is taken as ΔV tends to zero, that is

$$\gamma = \lim_{\Delta V \to 0} \frac{\Delta G}{\Delta V} = \frac{dG}{dV} \tag{2-3}$$

The weight is dependent on gravitational attraction and the gravitational acceleration **g** changes with location, so the specific weight depends on the local value of g. The relationship between γ and ρ can be deduced from Newton's second law.

$$G = mg$$
$$G/V = g(m/V)$$
$$\gamma = \rho g \tag{2-4}$$

The unit of γ is Newton per cubic meter (N/m^3). For water, the specific weight is $9.81 \times 10^3 \text{N/m}^3$; for air, it is 12.07N/m^3.

2.3.2 重度

重度（容重）是单位体积流体所具有的重量，用 γ 表示。

对均质流体，重度是流体的重量 G 与体积 V 的比值，因此

$$\gamma = \frac{G}{V} \tag{2-2}$$

对非均质流体，重度是指某一点的重度，某点附近非常小的体积内的流体重量 ΔG 与微小体积 ΔV 的比值，ΔV 趋近于 0，取其极限

$$\gamma = \lim_{\Delta V \to 0} \frac{\Delta G}{\Delta V} = \frac{dG}{dV} \tag{2-3}$$

重量随地球引力而发生变化，由于重力加速度 g 随地球上的位置不同而发生变化，重度 γ 随 g 而变化。γ 与 ρ 的关系可以由牛顿第二运动定律推求得到。

$$G = mg$$
$$G/V = g(m/V)$$
$$\gamma = \rho g \tag{2-4}$$

重度 γ 的单位是牛顿每立方米（N/m^3），水的重度是 $9.81 \times 10^3 \text{N/m}^3$，空气的重度是 12.07N/m^3。

2.4 Viscosity

Viscosity is the capability of a fluid to resist deformation and it is an inherent attribute of a fluid. The property that the shear stress comes into being in fluid when fluid flows is called the viscosity of a fluid.

2.4 黏性

黏性是流体抵抗变形的能力，它是流体的固有属性。流体运动时内部产生切应力的性质，叫作流体的黏性。

2.4.1 Newton's Law of Internal Friction

Because of the attractive forces between the molecules of a fluid and the momentum transfer brought from heat motion of molecules, the viscosity of the fluid comes into being and brings the internal friction also. In 1686, Newton brought forward the law of internal friction.

As shown in Figure 2-1, consider that there are two sufficiently large parallel plates, a small distance h apart, with general homogeneous fluid filling the space between. The lower board is stationary, while the upper one moves parallel to it with a properly constant velocity u due to a shear force F. The area A of the moving plate is so big that edge conditions of plates can be neglected.

Studying an infinitely thin fluid layer, the velocity of flow is u where the coordinate is y, and the velocity of flow is $u+\mathrm{d}u$ where the coordinate is $y+\mathrm{d}y$, thus it is obviously that the velocity gradient is $\dfrac{\mathrm{d}u}{\mathrm{d}y}$ in the thin layer with a thickness $\mathrm{d}y$. Newton considered that the value of internal friction T inside a fluid (namely the value of shear force F) has relations with the properties of the fluid, moreover, it is directly proportional to velocity gradient $\dfrac{\mathrm{d}u}{\mathrm{d}y}$ and contact area A, but has nothing to do with the pressure on the contact surface. In equation form

$$T = \pm \mu A \frac{\mathrm{d}u}{\mathrm{d}y} \qquad (2\text{-}5)$$

in which μ is the coefficient of viscosity (dynamic viscosity) and its unit is Pa·s. Dynamic viscosity has relations with the kind of fluid and temperature.

So we can express the shear stress τ by the following equation:

$$\tau = \frac{T}{A} = \pm \mu \frac{\mathrm{d}u}{\mathrm{d}y}$$

In Equation(2-5), the symbol $\pm$ is used to make T and τ positive. It is to say that when $\dfrac{\mathrm{d}u}{\mathrm{d}y}>0$ the positive sign is chosen, and when $\dfrac{\mathrm{d}u}{\mathrm{d}y}<0$ the minus sign is chosen. The physical meaning is that the shear stress is directly proportional to velocity gradient.

Consider a rectangular fluid parcel in the motional fluid deformed as a parallel quadrangle by a single shear stress τ during time $\mathrm{d}t$, as shown in Figure 2-2, we obtain

$$\frac{\mathrm{d}u}{\mathrm{d}y} = \frac{\mathrm{d}u}{\mathrm{d}y} \times \frac{\mathrm{d}t}{\mathrm{d}t} = \frac{\tan(\mathrm{d}\theta)}{\mathrm{d}t} \approx \frac{\mathrm{d}\theta}{\mathrm{d}t} \qquad (2\text{-}6)$$

So

$$\tau = \pm \mu \frac{du}{dy} = \pm \mu \frac{d\theta}{dt} \tag{2-7}$$

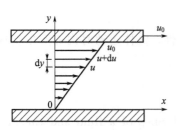

Figure 2-1 Velocity Distribution
图 2-1 速度分布

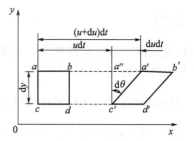

Figure 2-2 Velocity Gradient
图 2-2 速度梯度

Then we can know that the velocity gradient is equal to the shear strain rate of a fluid parcel and the shear stress is directly proportional to the shear strain rate in a fluid.

2.4.1 牛顿内摩擦定律

流体黏性是由于流体分子间的引力以及分子热运动产生动量交换而形成的并因此产生内摩擦力。1686 年牛顿提出牛顿内摩擦定律。

如图 2-1 所示,设有两个足够大且距离(h)很近的平行平板,中间充满一般的均质流体,下板固定,上板在切向力 F 作用下以不大的速度 u 做匀速直线运动。平板面积 A 足够大,以至于可忽略平板边缘的影响。

取无限薄的流体层进行研究,坐标为 y 处流速为 u,坐标为 $y+dy$ 处流速为 $u+du$,显然在厚度为 dy 的薄层中速度梯度为 $\frac{du}{dy}$,牛顿认为液层间内摩擦力 T 的大小(也就是切向力 F 的大小)与液体性质有关,并与速度梯度 $\frac{du}{dy}$ 和接触面积 A 成正比,而与接触面上的压力无关。即

$$T = \pm \mu A \frac{du}{dy} \tag{2-5}$$

式中,μ 为黏性系数(动力黏度),Pa·s,与流体种类、温度有关。

切应力可用下式表示:

$$\tau = \frac{T}{A} = \pm \mu \frac{du}{dy}$$

式(2-5) 中±号是为 T、τ 永为正值而设的,即当 $\frac{du}{dy} > 0$ 时取正号,当 $\frac{du}{dy} < 0$ 时取负号,其物理意义为切应力与速度梯度成正比。

在运动流体中取一矩形微元面,经 dt 时间变形运动成为一平行四边形(图 2-2),有:

$$\frac{du}{dy} = \frac{du}{dy} \times \frac{dt}{dt} = \frac{\tan(d\theta)}{dt} \approx \frac{d\theta}{dt} \tag{2-6}$$

故

$$\tau = \pm \mu \frac{du}{dy} = \pm \mu \frac{d\theta}{dt} \tag{2-7}$$

即速度梯度等于流体微团的剪切变形率,流体中的切应力与剪切变形率成正比。

【**Sample Problem 2-1**】 As shown in Figure 2-3, a plate is moving horizontally on oil. If the thickness δ of the oil film is 10mm and dynamic viscosity coefficient μ is 0.09807Pa·s, what is the required force F acting on unit area to drag the plate at a velocity u of 1m/s?

Solution: Because the thickness δ is very small, we can assume that u is linear with distance y.

$$\tau = \frac{T}{A} = \mu \frac{du}{dy} = \mu \frac{\Delta u}{\Delta y} = \mu \frac{u_2 - u_1}{y_2 - y_1}$$

$$\tau = \frac{T}{A} = \mu \frac{u_2 - u_1}{y_2 - y_1} = 0.09807 \frac{1-0}{0.01-0} = 9.8 \text{N/m}^2$$

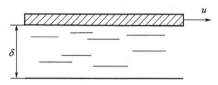

Figure 2-3　Figure for Sample Problem 2-1

图 2-3　例题 2-1 图

The force F acting on unit area is 9.8N/m^2.

【例题 2-1】 如图 2-3 所示，一平板在油面上做水平运动，已知平板运动速度 $u=1\text{m/s}$，板与固定边界的距离 $\delta=10\text{mm}$，油的动力黏度 $\mu=0.09807\text{Pa·s}$，则作用在平板单位面积上的拉力为多大？

解：因为油层厚度 δ 非常小，我们可以假设速度 u 与距离 y 呈线性变化。

$$\tau = \frac{T}{A} = \mu \frac{du}{dy} = \mu \frac{\Delta u}{\Delta y} = \mu \frac{u_2 - u_1}{y_2 - y_1}$$

$$\tau = \frac{T}{A} = \mu \frac{u_2 - u_1}{y_2 - y_1} = 0.09807 \frac{1-0}{0.01-0} = 9.8 \text{N/m}^2$$

作用在平板单位面积上的拉力为 9.8N/m^2。

2.4.2　Kinematic and Dynamic Viscosity Coefficient

During researches on the motion of fluids, the ratio of dynamic viscosity μ to density ρ, called kinematic viscosity coefficient, is used usually. It is expressed by ν, namely

$$\nu = \frac{\mu}{\rho} \tag{2-8}$$

where the unit of ν is m^2/s.

The physical meaning of μ is the shear stress under unit velocity gradient. So we can estimate the magnitude of viscosity of the same kind of fluid by μ.

The physical meaning of ν is the ratio of dynamic viscosity to density. If there is a great difference in the fluid densities, we should estimate the magnitude of viscosity according to ν instead of μ. The physical properties of water and air at standard atmospheric pressure are shown in Table 2-1 and Table 2-2.

2.4.2　运动黏性系数和动力黏性系数

在研究流体运动时，常常使用 μ 与密度 ρ 的比值，称为运动黏性系数（运动黏度），以 ν 表示，即

$$\nu = \frac{\mu}{\rho} \tag{2-8}$$

式中，运动黏性系数 ν 的单位是 m^2/s。

μ 的物理意义为单位速度梯度下的切应力，根据 μ 的大小可直接判断同种流体黏性的大小。ν 的物理意义为动力黏度与密度之比。如果流体密度相差很多，不能根据 μ 的大小判断黏性的大小，而应根据 ν 判断黏性的大小。表 2-1 和表 2-2 所示为标准大气压下水和空气的物理特性。

Table 2-1 Physical Properties of Water at Standard Atmospheric Pressure
表 2-1 标准大气压下水的物理特性

Temperature 温度/℃	Specific weight 重度 γ /(kN/m^3)	Density 密度 ρ/(kg/m^3)	Dynamic viscosity coefficient 动力黏度 $\mu \times 10^3$ /(N·s/m^2)	Kinematic viscosity coefficient 运动黏度 $\nu \times 10^6$/(m^2/s)	Surface tension 表面张力 σ/(N/m)	Absolute vapor pressure 绝对蒸气压 p_v/(kN/m^2)	Elastic modulus 弹性模量 $E \times 10^{-6}$ /(kN/m^2)
0	9.806	999.9	1.781	1.785	0.0756	0.16	2.02
5	9.807	1000.0	1.518	1.519	0.0749	0.87	2.06
10	9.804	999.7	1.307	1.306	0.0742	1.23	2.10
15	9.798	999.1	1.139	1.139	0.0735	1.70	2.15
20	9.789	998.2	1.002	1.003	0.0728	2.34	2.18
25	9.777	997.1	0.890	0.893	0.0720	3.17	2.22
30	9.764	995.7	0.798	0.800	0.0712	4.24	2.25
40	9.730	992.2	0.653	0.658	0.0696	7.38	2.28
50	9.689	988.0	0.547	0.553	0.0679	12.33	2.29
60	9.642	983.2	0.466	0.474	0.0662	19.92	2.28
70	9.584	977.8	0.404	0.413	0.0644	31.16	2.25
80	9.530	971.8	0.354	0.364	0.0626	47.34	2.20
90	9.466	965.3	0.315	0.326	0.0608	70.10	2.14
100	9.399	958.4	0.282	0.294	0.0589	101.33	2.07

Table 2-2 Physical Properties of Air at Standard Atmospheric Pressure
表 2-2 标准大气压下空气的物理特性

Temperature 温度/℃	Density 密度 ρ/(kg/m^3)	Specific weight 重度 γ/(kN/m^3)	Dynamic viscosity coefficient 动力黏度 $\mu \times 10^5$/(N·s/m^2)	Kinematic viscosity coefficient 运动黏度 $\nu \times 10^5$/(m^2/s)
−40	1.515	14.86	1.49	0.98
−20	1.395	13.68	1.61	1.15
0	1.293	12.68	1.71	1.32
10	1.248	12.24	1.76	1.41
20	1.205	11.82	1.81	1.50
30	1.165	11.43	1.86	1.60
40	1.128	11.06	1.90	1.68
60	1.006	10.40	2.00	1.87
80	1.000	9.81	2.09	2.09
100	0.946	9.28	2.18	2.31
200	0.747	7.33	2.58	3.45

【Sample Problem 2-2】 At the same temperature, $\mu_{\text{water}} > \mu_{\text{air}}$. Try to demonstrate that at 20℃ for water and air which one is easier to flow.

Solution: At 20℃ the following data are gotten from the physical properties tables of water and air.

$$\rho_{\text{water}} = 998.2 \text{kg/m}^3, \quad \rho_{\text{air}} = 1.205 \text{kg/m}^3$$

$$\mu_{\text{water}} = 1.002 \times 10^{-3} \text{Pa} \cdot \text{s}, \quad \mu_{\text{air}} = 1.81 \times 10^{-5} \text{Pa} \cdot \text{s}$$

$$\nu_{\text{water}} = 1.003 \times 10^{-6} \text{m}^2/\text{s}, \quad \nu_{\text{air}} = 15.0 \times 10^{-6} \text{m}^2/\text{s}$$

$$\frac{\mu_{\text{water}}}{\mu_{\text{air}}} = \frac{1.002 \times 10^{-3}}{1.81 \times 10^{-5}} = 55.36 \text{times}$$

$$\frac{\rho_{\text{water}}}{\rho_{\text{air}}} = \frac{998.2}{1.205} = 828.38 \text{times}$$

$$\frac{\nu_{\text{air}}}{\nu_{\text{water}}} = \frac{15.0 \times 10^{-6}}{1.003 \times 10^{-6}} = 14.96 \text{times}$$

At the same temperature $\nu_{\text{air}} > \nu_{\text{water}}$. Compared with water, the air is hard to flow. The example shows that we should estimate the fluidness of a fluid from ν instead of μ directly at the same temperature, because ν excludes the influence of fluid density and only keeps the property parameters of motion. This is the meaning of kinematic viscosity coefficient.

【例题 2-2】在相同温度下 $\mu_水 > \mu_{空气}$，试论证在 20℃时，水和空气相比，哪种流体易于流动。

解：在 20℃时，从水和空气物理特性表中查取

$$\rho_水 = 998.2 \text{kg/m}^3, \quad \rho_{空气} = 1.205 \text{kg/m}^3$$

$$\mu_水 = 1.002 \times 10^{-3} \text{Pa} \cdot \text{s}, \quad \mu_{空气} = 1.81 \times 10^{-5} \text{Pa} \cdot \text{s}$$

$$\nu_水 = 1.003 \times 10^{-6} \text{m}^2/\text{s}, \quad \nu_{空气} = 15.0 \times 10^{-6} \text{m}^2/\text{s}$$

$$\frac{\mu_水}{\mu_{空气}} = \frac{1.002 \times 10^{-3}}{1.81 \times 10^{-5}} = 55.36 \text{（倍）}$$

$$\frac{\rho_水}{\rho_{空气}} = \frac{998.2}{1.205} = 828.38 \text{（倍）}$$

$$\frac{\nu_{空气}}{\nu_水} = \frac{15.0 \times 10^{-6}}{1.003 \times 10^{-6}} = 14.96 \text{（倍）}$$

在相同温度下，$\nu_{空气} > \nu_水$，与水相比，空气不易流动。此例题说明在相同温度下，只从 μ 值的大小不能直接判断流体的流动性，而应根据 ν 值的大小直接判别流体的流动性，因为 ν 值排除了流体密度的影响，只保留其运动特征参数。这也是引入运动黏度的意义所在。

2.4.3 Relationship between Viscosity and Temperature

Generally speaking, the viscosity of a fluid increases only weakly with pressure, so the effect of pressure can be neglected normally.

Temperature, however, has a strong effect on the viscosity of a fluid. The gas viscosity is caused mainly by the momentum transfer coming from the thermal movement of molecules. So when the temperature goes up, the thermal movement of molecules becomes strong and viscosity augments. The liquid viscosity is caused mainly by the attractive forces between molecules. So when the temperature goes up, the momentum of molecules enlarges, attractive forces decrease and viscosity declines.

2.4.3 黏性与温度的关系

一般来说，增加压力，流体的黏性变化很小，因而一般不考虑压力的影响。

然而，流体的黏性受温度的影响很大。气体的黏性主要是由分子热运动产生的动量交换引起的，因此当温度升高时，分子热运动加剧，黏性增大。液体的黏性主要是由分子之间的引力产生的，当温度升高时，分子动量增加，引力减小，黏性下降。

2.4.4 Newtonian and Non-Newtonian Fluids

Fluids obeying Newton's law of viscosity and for which μ has a constant value are known as

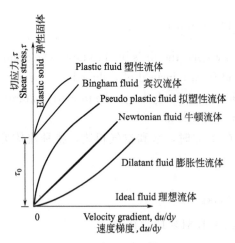

Figure 2-4 Newtonian Fluids and Non-Newtonian Fluids

图 2-4 牛顿流体与非牛顿流体

Newtonian fluids. The shear stress of a Newtonian fluid is linearly related to velocity gradient. Fluids which do not obey Newton's law of viscosity are known as non-Newtonian fluids. Newtonian fluids and non-Newtonian fluids are shown in Figure 2-4.

2.4.4 牛顿流体与非牛顿流体

遵循牛顿内摩擦定律,并且 μ 为常数的流体称为牛顿流体,它的剪切应力与速度梯度呈线性关系。不遵循牛顿内摩擦定律的流体称为非牛顿流体。牛顿流体与非牛顿流体如图 2-4 所示。

2.4.5 Ideal Fluids and Real Fluids

An ideal fluid is usually defined as a fluid in which there is no friction. Its viscosity is zero. In fact all fluids have viscosity. The purpose of putting forward the concept of ideal fluid is to research the fluid movement rules and simplify the derivation of the theory questions greatly.

In a real fluid, either liquid or gas, shearing forces always develop whenever there is motion relative to a body, thus creating fluid friction, and these forces will hinder the motion of one particle past another.

2.4.5 理想流体和实际流体

理想流体通常被定义为没有摩擦的流体,它的黏性系数是 0。实际上,一切流体都具有黏性,提出理想流体的概念的目的在于研究流体运动规律时,对理论方程的推导大为简化。

在实际流体(不论气体或者液体)中,只要内部存在相对运动,就会有剪切力,产生摩擦,阻碍相对运动。

2.5 Compressibility and Expansivity

The relative density, density and specific volume of a fluid change with temperature and pressure, due to the space between the molecules inside a fluid. With the increase of pressure, the distances between molecules diminish and the volume is compressed. The distances between molecules increase and the volume expands with the increase of temperature. All fluids have the properties of compressibility and expansivity.

2.5 压缩性和膨胀性

流体相对密度、密度、比体积随温度与压强变化,其原因是流体内部分子间存在间隙。压强增大,分子间距减小,体积压缩;温度升高,分子间距增大,体积膨胀。流体都具有这种可压缩、能膨胀的性质。

2.5.1 Compressibility

The property that the volume of a fluid decreases under pressure at constant temperature is called compressibility. The degree of compressibility is represented by volume compression coefficient β_p.

$$\beta_p = -\frac{\dfrac{dV}{V}}{dp} = \frac{1}{\rho}\frac{d\rho}{dp} \tag{2-9}$$

where V—original volume, m^3;

dV—change in volume, m^3;

dp—change in pressure, Pa;

β_p—the volume compression coefficient, Pa^{-1}.

The physical meaning of β_p is the relative rate of volume change of a fluid with increasing per unit pressure under constant temperature.

The reciprocal of β_p is called elastic modulus of a fluid, which is expressed by E.

$$E = \frac{1}{\beta_p} \tag{2-10}$$

The compression rate of a gas under the condition of constant temperature can be obtained from the equation of state (let $T=C$)

$$\beta_p = -\frac{1}{V}\frac{d}{dp}\left(\frac{mR_g T}{p}\right) = -\frac{mR_g T}{V}\left(-\frac{1}{p^2}\right) = \frac{1}{p} \tag{2-11}$$

where m—the mass of the gas, kg;

R_g—a gas constant.

According to Equation (2-11), β_p is inversely proportional to p.

2.5.1 压缩性

在温度不变的条件下，流体在压力作用下体积缩小的性质称为压缩性。压缩性的大小用体积压缩系数 β_p 表示，即

$$\beta_p = -\frac{\dfrac{dV}{V}}{dp} = \frac{1}{\rho}\frac{d\rho}{dp} \tag{2-9}$$

式中，V 为原有体积，m^3；dV 为体积改变量，m^3；dp 为压强改变量，Pa；β_p 为体积压缩系数，Pa^{-1}。

β_p 的物理意义为当温度不变时每增加单位压强所产生的流体体积相对变化率。

流体压缩系数的倒数称为流体的弹性模量，用 E 表示，即

$$E = \frac{1}{\beta_p} \tag{2-10}$$

气体的等温压缩率亦可由气体状态方程（令 $T=C$）求得

$$\beta_p = -\frac{1}{V}\frac{d}{dp}\left(\frac{mR_g T}{p}\right) = -\frac{mR_g T}{V}\left(-\frac{1}{p^2}\right) = \frac{1}{p} \tag{2-11}$$

式中，m 为气体的质量，kg；R_g 为气体常数。

由式(2-11)可见，β_p 与 p 成反比。

2.5.2 Expansivity

The property that the volume of a fluid increases with the increase of temperature under constant pressure is called expansivity.

The degree of expansivity is represented by volume expansion coefficient β_t.

$$\beta_t = \frac{1}{V}\frac{dV}{dT} = -\frac{1}{\rho}\frac{d\rho}{dT} \tag{2-12}$$

where dT — the temperature increment, K;

β_t — the volume expansion coefficient, K^{-1};

$\dfrac{dV}{dT}$ — the rate of change of volume when temperature changes, m^3/K.

The physical meaning of β_t is the relative rate of volume change of a fluid with increasing per unit temperature under constant pressure.

The compression coefficient of water is shown in Table 2-3, the expansion coefficient of water is shown in Table 2-4, specific weight and density of water is shown in Table 2-5.

2.5.2 膨胀性

在压力不变的条件下,温度升高,流体体积增大的性质称为膨胀性。膨胀性的大小用体积膨胀系数 β_t 表示,即

$$\beta_t = \frac{1}{V}\frac{dV}{dT} = -\frac{1}{\rho}\frac{d\rho}{dT} \tag{2-12}$$

式中,dT 为温度变化量,K;β_t 为体积膨胀系数,K^{-1};dV/dT 为相对于温度变化的体积变化率,m^3/K。

β_t 的物理意义为当压强不变时每增加单位温度所产生的流体体积相对变化率。

水的压缩系数见表 2-3,水的膨胀系数见表 2-4,水的重度和密度见表 2-5。

Table 2-3 The Compression Coefficient of Water

表 2-3 水的压缩系数

Pressure 压强/at	5	10	20	40	80
$\beta_p \times 10^{-9}/(m^2/N)$	0.538	0.536	0.531	0.528	0.515

注:1at=98066.5Pa。

Table 2-4 The Expansion Coefficient of Water

表 2-4 水的膨胀系数

Temperature 温度/℃	1~10	10~20	40~50	60~70	90~100
$\beta_t \times 10^{-4}/℃^{-1}$	0.14	0.15	0.42	0.55	0.72

Table 2-5 Specific Weight and Density of Water

表 2-5 水的重度和密度

Temperature 温度/℃	Specific weight 重度/(kN/m³)	Density 密度/(kg/m³)	Temperature 温度/℃	Specific weight 重度/(kN/m³)	Density 密度/(kg/m³)	Temperature 温度/℃	Specific weight 重度/(kN/m³)	Density 密度/(kg/m³)
0	9.806	999.9	10	9.804	999.7	60	9.645	983.2
1	9.806	999.9	15	9.798	999.1	65	9.617	980.6
2	9.807	1000.0	20	9.789	998.2	70	9.590	977.8
3	9.807	1000.0	25	9.778	997.1	75	9.561	974.9
4	9.807	1000.0	30	9.764	995.7	80	9.529	971.8
5	9.807	1000.0	35	9.749	994.1	85	9.500	968.7
6	9.807	1000.0	40	9.731	992.2	90	9.467	965.3
7	9.806	999.9	45	9.710	990.2	95	9.433	961.9
8	9.806	999.9	50	9.690	988.1	100	9.339	958.4
9	9.805	999.8	55	9.657	985.7		—	—

【Sample Problem 2-3】 Calculate the relative change rate of density when the water pressure is increased from 5at to 10at under normal temperature.

Solution: This problem is the application of compression coefficient equation of a liquid under normal temperature.

Solution 1:
$$\beta_p = \frac{1}{\rho} \times \frac{d\rho}{dp}$$

Thus,
$$\frac{d\rho}{\rho} = \beta_p dp = 0.538 \times 10^{-9} \times (10-5) \times 9.81 \times 10^4 = 0.0264\%$$

Solution 2: Integrate the equation
$$\frac{d\rho}{\rho} = \beta_p dp$$

Then
$$\ln\rho - \ln\rho_0 = \beta_p (p_2 - p_1)$$

Thus
$$\frac{\rho}{\rho_0} = \exp[\beta_p (p_2 - p_1)]$$
$$= \exp[0.538 \times 10^{-9} \times (10-5) \times 98100] = 1.000264$$
$$\frac{\rho - \rho_0}{\rho_0} = 0.0264\%$$

So the relative change rate of density is 0.0264%.

【例题 2-3】 水在常温下由 5at 增加到 10at 时，求水的密度的相对变化率。

解：此题是液体在常温下压缩系数公式的应用。

解法一：
$$\beta_p = \frac{1}{\rho} \times \frac{d\rho}{dp}$$

则
$$\frac{d\rho}{\rho} = \beta_p dp = 0.538 \times 10^{-9} \times (10-5) \times 9.81 \times 10^4 = 0.0264\%$$

解法二：对 $\frac{d\rho}{\rho} = \beta_p dp$ 进行积分

有
$$\ln\rho - \ln\rho_0 = \beta_p(p_2 - p_1)$$

则
$$\frac{\rho}{\rho_0} = \exp[\beta_p(p_2 - p_1)]$$
$$= \exp[0.538 \times 10^{-9} \times (10-5) \times 98100] = 1.000264$$
$$\frac{\rho - \rho_0}{\rho_0} = 0.0264\%$$

故水密度的相对变化率为 0.0264%。

2.5.3 Incompressible Fluids

The fluid, whether liquid or gas, is compressible, and the volume V of a given mass will be reduced to $V - dV$ when a force is exerted uniformly all over its surface. Fluid mechanics usually deals with both incompressible and compressible fluids, that is, liquids and gases with either constant or variable density. If the change in density with pressure is so small as to be negligible, that is an incompressible fluid. This is usually the case with liquids. We may also consider gases to be incompressible when the pressure variation is small compared with the absolute pressure.

2.5.3 不可压缩流体

当外力均匀作用在流体的外表面时，不论液体或者气体都会被压缩，体积由 V 减少到 $V-dV$。在流体力学中，根据密度的变化，流体通常分为不可压缩和可压缩流体。如果密度变化很小，可以忽略，就称为不可压缩流体。液体通常被认为是不可压缩流体。对于气体来说，如果压强的变化量与绝对压强比起来很小，也可以认为是不可压缩流体。

2.6 Surface Tension

Because of the attractive forces between molecules, the tiny tension exerted on the free surface of a fluid is known as surface tension. As show in Figure 2-5, a fine glass tube with two open ends up is put straightly into a liquid. The liquid in the tube will rise or depress under the action of surface tension. This phenomenon is called capillarity.

The weight of the liquid column equals to the vertical component of the accessional press due to the surface tension, i.e.,

$$\frac{\pi}{4}d^2 h\gamma = \pi d\sigma \cos\alpha \qquad (2\text{-}13)$$

Thus
$$h = \frac{4\sigma\cos\alpha}{\gamma d} \qquad (2\text{-}14)$$

where h — the rise of liquid surface, m;
d — the diameter of glass tube, m;
γ — specific weight of liquid, N/m³;
σ — coefficient of surface tension, N/m;
α — contact angle, (°).

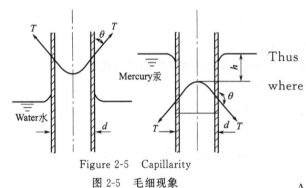

Figure 2-5 Capillarity

图 2-5 毛细现象

According to experiments, at 20℃, the contact angle of water and glass $\alpha = 3°\sim 9°$ and the coefficient of surface tension $\sigma = 0.0728\text{N/m}$; the contact angle of mercury and glass $\alpha = 139°\sim 140°$ and the coefficient of surface tension $\sigma = 0.51\text{N/m}$.

2.6 表面张力

由于分子间存在引力，在液体自由表面上能承受的微小张力称为表面张力。如图 2-5 所示，将两端开口的细玻璃管竖立在液体中，在表面张力的作用下，管中液体将上升或下降一个高度，该现象称为毛细现象。

由于液体柱的重力与表面张力产生的附加压力在垂直方向上的分力平衡，所以有：

$$\frac{\pi}{4}d^2 h\gamma = \pi d\sigma \cos\alpha \qquad (2\text{-}13)$$

则
$$h = \frac{4\sigma\cos\alpha}{\gamma d} \qquad (2\text{-}14)$$

式中，h 为液面上升高度，m；d 为玻璃管直径，m；γ 为液体重度，N/m³；σ 为表面张力系数，N/m；α 为接触角，(°)。

实验测得：20℃时水与玻璃的接触角 $\alpha=3°\sim9°$，表面张力系数 $\sigma=0.0728\text{N/m}$；汞与玻璃的接触角 $\alpha=139°\sim140°$，$\sigma=0.51\text{N/m}$。

2.7 Vapor Pressure

Liquids tend to evaporate, which they do by projecting molecules into the space above their surfaces. If this happens in a confined space, equilibrium will be reached. The partial pressure exerted by the vaporizing molecules is known as vapor pressure. Molecular activity increases with increasing temperature and decreasing pressure. At any given temperature, if the vapor pressure is equal to the pressure on the liquid surface, a boiling results. Reduce the pressure, the boiling can result at a temperature below the boiling point at atmospheric pressure. For example, if the pressure is reduced to 12.33kPa, the water boils at 50℃.

The vapor bubbles form when the pressure falls below the vapor pressure in many situations involving the flow of liquids. This phenomenon is called cavitation. The growth and decay of the vapor bubbles will affect the operating performances of hydraulic machinery such as pumps, propellers and turbines. When the cloud of bubbles collapses suddenly, the very large force of liquids will hit the solid surface. The impact of collapsing bubbles can cause local erosion of metal surfaces.

2.7 蒸气压

液体的蒸发是由于分子会从液体的表面逸出到空气中，假如液体上方的空间是有限的，蒸发可以达到平衡，气态分子产生的压强就称为蒸气压。随着温度的升高及压强的减小，分子运动将加剧。在一定的温度下，当蒸气压达到液体表面的压强时，沸腾就发生了。通过降低压强，沸腾可以在低于大气压下沸点的某一温度下发生。例如，当气压减小到 12.33kPa 时，水在 50℃就沸腾了。

液体内部的压强低于蒸气压时，气泡将会产生，这种现象称为气穴现象。气泡的产生和破灭将会影响水泵、螺旋桨及水轮机等水力机械的操作运行。当一团气泡忽然破灭时，来自液体的巨大的冲击力就会打在固体的表面，这种气泡破灭将会导致金属表面的局部腐蚀。

2.8 Forces on a Fluid

Each fluid particle is acted by all kinds of forces no matter it keeps moving or balance. According to the behaviors of the forces, there are two classes of forces: body forces and surface forces.

2.8 作用在流体上的力

流体的每一质点无论处于运动还是平衡状态，都受到各种力的作用，按力的表现形式可分为质量力和表面力两类。

2.8.1 Body Forces

A force, which is directly proportional to the mass of a fluid parcel and acts on the mass center, is called body force. Body forces include gravitation and inertial force.

Gravitation $\qquad \vec{G} = m\vec{g}$

Inertial force of linear motion $\qquad \vec{F} = m\vec{a}$

The body force acting on unit mass fluid is called unit body force, and is expressed as follows:

$$\vec{f} = \lim_{\Delta m \to 0} \frac{\Delta \vec{F}}{\Delta m} = \frac{d\vec{F}}{dm} \tag{2-15}$$

$$\vec{f} = f_x \vec{i} + f_y \vec{j} + f_z \vec{k} \tag{2-16}$$

$$d\vec{F} = dm\vec{f} = dm\ (f_x \vec{i} + f_y \vec{j} + f_z \vec{k}) \tag{2-17}$$

where $\quad \Delta m$—the mass of the fluid parcel;

$\Delta \vec{F}$—the body force acting on the parcel;

f_x, f_y and f_z—the projections of the unit body force in the x, y and z directions, respectively.

2.8.1 质量力

与流体微团质量成正比并且集中作用在微团质量中心上的力称为质量力,质量力包括重力和惯性力。

重力 $\qquad \vec{G} = m\vec{g}$

直线运动惯性力 $\qquad \vec{F} = m\vec{a}$

单位质量流体所受的质量力称为单位质量力,记作

$$\vec{f} = \lim_{\Delta m \to 0} \frac{\Delta \vec{F}}{\Delta m} = \frac{d\vec{F}}{dm} \tag{2-15}$$

$$\vec{f} = f_x \vec{i} + f_y \vec{j} + f_z \vec{k} \tag{2-16}$$

$$d\vec{F} = dm\vec{f} = dm\ (f_x \vec{i} + f_y \vec{j} + f_z \vec{k}) \tag{2-17}$$

式中,Δm 为流体微元体的质量;$\Delta \vec{F}$ 为作用在该微元体上的质量力;f_x、f_y 和 f_z 分别为单位质量力在 x、y、z 轴的投影。

2.8.2 Surface Forces

A force, which is directly proportional to surface area of a fluid and distributes on the fluid surface, is called surface force. According to the acting direction, it can be classified as the pressure force along the internal normal direction of the surface and the friction force along the tangential direction of the surface.

Element of area ΔA lies in the fluid parcel. Suppose the tiny pressure force acting on ΔA is $\Delta \vec{F}$ and the tiny shear force is $\Delta \vec{T}$. Then the compressive stress at a point is

$$\vec{p} = \lim_{\Delta A \to 0} \frac{\Delta \vec{F}}{\Delta A} \tag{2-18}$$

So the pressure force is

$$\vec{F} = \vec{p}A \tag{2-19}$$

The shear stress at a point is

$$\vec{\tau} = \lim_{\Delta A \to 0} \frac{\Delta \vec{T}}{\Delta A} \tag{2-20}$$

So the shear force is

$$\vec{T} = \vec{\tau}A \tag{2-21}$$

2.8.2 表面力

大小与流体表面积成正比，而且分散作用在流体表面上的力称为表面力，按其作用方向分为沿表面内法线方向的压力和沿表面切向的摩擦力。

在流体微团上取微元面积 ΔA，设作用在 ΔA 上的微小压力为 $\Delta \vec{F}$，微小切向力为 $\Delta \vec{T}$。则各点处的压应力为：

$$\vec{p} = \lim_{\Delta A \to 0} \frac{\Delta \vec{F}}{\Delta A} \tag{2-18}$$

故表面上的压力为：

$$\vec{F} = \vec{p}A \tag{2-19}$$

各点处的切应力为：

$$\vec{\tau} = \lim_{\Delta A \to 0} \frac{\Delta \vec{T}}{\Delta A} \tag{2-20}$$

故表面上的切向力为：

$$\vec{T} = \vec{\tau}A \tag{2-21}$$

Exercises 习题

2-1 A block of weight W slides down an inclined plane while lubricated by a thin film of oil, as shown in the figure. The contact area of the block and the plane is A, the thickness of the film is h. Assuming a linear velocity distribution in the film, derive an expression for the "terminal" velocity v of the block.

2-1 一块质量为 W 的物块沿着一个表面有薄层石油的斜面下滑，如图所示。物体和斜面的接触面积为 A，液膜厚度为 h。假设薄膜中的速度是线性分布的，试求物块到达斜面末端时速度 v 的表达式。

2-2 As shown in the figure, oil of absolute viscosity μ fills the small gap of thickness Y. Neglecting fluid stress exerted on the circular underside, obtain an expression for the torque T required to rotate the truncated cone at constant speed ω.

2-2 如图所示，用绝对黏度为 μ 的石油填补厚度为 Y 的间隙，忽略流体应力对圆形底面的作用，求能使截锥体以 ω 匀速旋转运动的扭矩 T 的表达式。

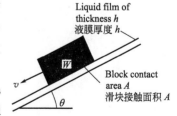

Figure for Exercise 2-1

习题 2-1 图

2-3 A flat plate 200mm×800mm slides on oil ($\mu = 0.85$N·s/m²) over a large plane surface. What force F is required to drag the plate at a velocity v of 1.0m/s, if the thickness δ of the separating oil film is 1.0mm?

2-3 在一块面积很大的平面上有一层绝对黏度 $\mu=0.85\text{N}\cdot\text{s}/\text{m}^2$ 的油，上面有一块规格为 200mm×800mm 的平板，如图所示，如果板和平面之间的距离 δ 为 1.0mm，施加一个力 F，使该平板以 1.0m/s 的速度移动，求这个力为多大。

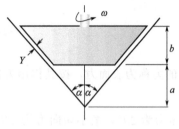

Figure for Exercise 2-2

习题 2-2 图

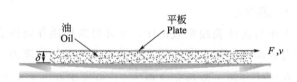

Figure for Exercise 2-3

习题 2-3 图

2-4 The original state of air is $p_0=101.3\text{kPa}$ and $t_0=15\text{℃}$. After adiabatic compression in the cylinder, the volume of the air decreases by half. What is the temperature and pressure of the end state?

2-4 空气初始状态为 $p_0=101.3\text{kPa}$、$t_0=15\text{℃}$，在汽缸内绝热压缩后体积减小了一半，求终态温度和压强。

2-5 In a heating system there is a dilatation water tank. The whole volume of the water in the system is 8m³. The largest temperature rise is 50℃ and the expansion coefficient is $\beta_t=0.005\text{℃}^{-1}$, what is the smallest volume of the water bank?

2-5 采暖系统在顶部设一膨胀水箱，系统内水的总体积为 8m³，最大温升为 50℃，膨胀系数 $\beta_t=0.005\text{℃}^{-1}$，求该水箱的最小容积。

Chapter 3　Fluid Statics
第3章　流体静力学

Fluid statics researches the mechanics rules and applications of equilibrium fluids. The static state can be described as that there are no relative motions between macroscopic particles and a relative equilibrium reaches. The static states of fluids include two cases. One is absolute static state, where the fluids have no relative motion to the earth. The other is relative static state, where the fluids have relative motion to the earth, but there are no relative motions between fluid particles.

There are no relative motions among the equilibrium fluids. In fluids there are no shear stresses; hence as surface forces, only normal static pressures are present.

The main contents of this chapter are the distribution rules of pressure in a static fluid and the total pressure force acting on the fixed wall. Moreover, on the basis of these, some engineering questions are solved.

流体静力学研究平衡流体的力学规律及其在工程上的应用。这里的静止是指流体宏观质点之间没有相对运动，达到了相对平衡。流体的静止状态包括两种情况：一种是流体对地球无相对运动，叫绝对静止；另一种是流体整体对地球有相对运动，但流体各质点之间没有相对运动，叫相对静止。

平衡流体相互之间没有相对运动，流体内不存在切向应力，作用在流体上的表面力只有法向的静压强。

本章的主要任务为研究流体静压强在空间的分布规律及平衡流体作用在固定壁面上的总压力等，并在此基础上解决一些工程实际问题。

3.1　Static Pressure and Its Characters

The pressure at a point in the equilibrium fluid is called static pressure in a fluid. It is expressed by the following equation:

$$p = \lim_{\Delta A \to 0} \frac{\Delta P}{\Delta A}$$

where　ΔA—the area of the element;
　　　ΔP—the total pressure force acting on ΔA.

The vector expression of the static pressure in a fluid on the surface of the element is

$$\mathrm{d}\vec{P} = -p\,\mathrm{d}\vec{A} \tag{3-1}$$

Both the magnitudes and directions have some relations to the compression surface, and the minus shows that the direction of static pressure goes along the inner normal direction of the compression surface.

3.1 静压强及其特性

平衡流体中的某一点的压强称为流体静压强，表示为

$$p = \lim_{\Delta A \to 0} \frac{\Delta P}{\Delta A}$$

式中，ΔA 为微元面积；ΔP 为作用在 ΔA 表面上的总压力大小。

微元表面上的流体静压力矢量表达式为

$$\mathrm{d}\vec{P} = -p\,\mathrm{d}\vec{A} \tag{3-1}$$

其大小与方向均与受压面有关。负号说明流体静压力的方向是沿受压面的内法线方向。

The characters of static pressure are as follows.

(1) The direction always goes along the inner normal of the compression surface Incise the static fluid into two parts with an arbitrary plane, just as shown in Figure 3-1. Take the shadow part as partition, if the direction of the static pressure p at a certain point on incisory plane doesn't go along the normal direction but arbitrary, then p can be decomposed into tangent component τ and normal component p_n. The static fluid does not undergo shear stress and pulling force or else the equilibrium will be destroyed. So the only direction of static pressure is consistent with the normal direction on the acting surface.

流体静压强的特性如下。

(1) 静压强方向永远沿着作用面内法线方向 用任意一个平面将静止流体切割为两部分，见图 3-1，取阴影部分为隔离体，如果切割平面上某一点处静压力方向不是法线方向而是任意方向的，则 p 可分解为切向分量 τ 和法向分量 p_n。静止流体既不承受切应力，也不承受拉力，否则平衡将被破坏，所以静压力唯一可能的方向就是和作用面内法线方向一致。

(2) Pressure at a point is the same in any direction In the equilibrium fluid, a very small wedge-shaped element $OABC$ of size $\mathrm{d}x$ by $\mathrm{d}y$ by $\mathrm{d}z$ is shown in Figure 3-2.

Assume that the pressure at a random point of each surface of the wedge-shaped element is expressed by p_x, p_y, p_z and p_n respectively. Then the surface forces acting on the wedge-shaped element are

$$P_x = \frac{1}{2} p_x \mathrm{d}y \mathrm{d}z$$

$$P_y = \frac{1}{2} p_y \mathrm{d}x \mathrm{d}z$$

$$P_z = \frac{1}{2} p_z \mathrm{d}x \mathrm{d}y$$

$$P_n = p_n \mathrm{d}S$$

where $\mathrm{d}S$ —— the area of $\triangle ABC$.

Thus

$$\begin{aligned}\mathrm{d}\vec{P} &= [p_x \frac{1}{2}\mathrm{d}y\mathrm{d}z - p_n \mathrm{d}S\cos(n,x)]\vec{i} + [p_y \frac{1}{2}\mathrm{d}x\mathrm{d}z - p_n \mathrm{d}S\cos(n,y)]\vec{j} + \\ &\quad [p_z \frac{1}{2}\mathrm{d}x\mathrm{d}y - p_n \mathrm{d}S\cos(n,z)]\vec{k} \\ &= (p_x - p_n)\frac{1}{2}\mathrm{d}y\mathrm{d}z\vec{i} + (p_y - p_n)\frac{1}{2}\mathrm{d}x\mathrm{d}z\vec{j} + (p_z - p_n)\frac{1}{2}\mathrm{d}x\mathrm{d}y\vec{k}\end{aligned}$$

The mass force on the element is

$$d\vec{F}_m = \rho dV \vec{f}_m = \rho \frac{1}{6} dx\,dy\,dz\,(f_x\vec{i} + f_y\vec{j} + f_z\vec{k})$$

Since the fluid is in equilibrium, $\sum \vec{F} = 0$.

Simplify the equation and we can obtain:

$$p_x - p_n + f_x \rho \frac{1}{3} dx = 0$$

$$p_y - p_n + f_y \rho \frac{1}{3} dy = 0$$

$$p_z - p_n + f_z \rho \frac{1}{3} dz = 0$$

When dx, dy and dz go to zero, the wedge-shaped element will lessen to a point O. The pressures p_x, p_y, p_z and p_n are equal to the static pressure in a fluid at point O in all directions. So we can obtain:

$$p_x = p_y = p_z = p_n$$

Normally, the static pressures in a fluid at different points are different to each other. That is to say, the static pressure in a fluid is a continuous function of space coordinates. Namely

$$p = p(x, y, z)$$

（2）静止流体中任何一点上各个方向的静压强大小相等，与作用面方位无关　在平衡流体中任取边长为 dx、dy、dz 的微元四面体 $OABC$，如图 3-2 所示。

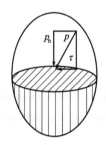

Figure 3-1　The Direction of Static Pressure　　Figure 3-2　Tetrahedron Element of the Equilibrium Fluid
　　图 3-1　静压强的方向　　　　　　　　　图 3-2　平衡流体中的微元四面体

设四面体每个面上任意一点的压强分别用 p_x、p_y、p_z 及 p_n 表示，则作用在微元四面体表面上的力为：

$$P_x = \frac{1}{2} p_x dy\,dz$$

$$P_y = \frac{1}{2} p_y dx\,dz$$

$$P_z = \frac{1}{2} p_z dx\,dy$$

$$P_n = p_n dS$$

式中，dS 为 $\triangle ABC$ 的面积。

因此，

$$d\vec{P} = [p_x \frac{1}{2}dydz - p_n dS\cos(n,x)]\vec{i} + [p_y \frac{1}{2}dxdz - p_n dS\cos(n,y)]\vec{j} + [p_z \frac{1}{2}dxdy - p_n dS\cos(n,z)]\vec{k}$$

$$= (p_x - p_n)\frac{1}{2}dydz\vec{i} + (p_y - p_n)\frac{1}{2}dxdz\vec{j} + (p_z - p_n)\frac{1}{2}dxdy\vec{k}$$

微元流体上的质量力为：

$$d\vec{F}_m = \rho dV \vec{f}_m = \rho \frac{1}{6} dxdydz \ (f_x\vec{i} + f_y\vec{j} + f_z\vec{k})$$

流体处于平衡状态，根据 $\Sigma \vec{F} = 0$，简化后有：

$$p_x - p_n + f_x \rho \frac{1}{3} dx = 0$$

$$p_y - p_n + f_y \rho \frac{1}{3} dy = 0$$

$$p_z - p_n + f_z \rho \frac{1}{3} dz = 0$$

dx、dy、dz 趋于零时，四面体缩到 O 点，其上任何一点的压强 p_x、p_y、p_z 和 p_n 就变成 O 点上各个方向的流体静压强，于是得到

$$p_x = p_y = p_z = p_n$$

不同空间点的流体静压强，一般是各不相同的，即流体静压强是空间坐标的连续函数。

$$p = p(x,y,z)$$

3.2　Differential Equation of Fluid Equilibrium

3.2.1　Euler's Equilibrium Equation

A rectangular parallelepiped element of fluid is shown in Figure 3-3. The dimensions of the element are dx, dy and dz and its volume is $dxdydz$. The density is ρ and the pressure is $p(x,y,z)$ at a point.

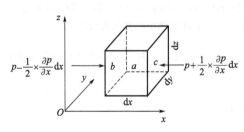

Figure 3-3　Hexahedral Element of Fluid
图 3-3　微元六面体

The forces acting on an element of the static fluid consist of body forces and surface forces.

(1) Body forces　Body force = body force unit mass × mass = body force unit mass × density × volume

The components F_x, F_y, F_z of body forces in the x, y, z direction are as follows:

$$F_x = f_x \rho dxdydz$$
$$F_y = f_y \rho dxdydz$$
$$F_z = f_z \rho dxdydz$$

(2) Surface forces　Since the fluid is at rest, there are no shear forces on the surfaces of the element.

Surface force＝Pressure × Area of the surface

The pressure at the center of the element is $p(x, y, z)$, so the force exerted at b point of the side normal to the x axis is approximately

$$P_b = p_b \mathrm{d}y\mathrm{d}z = \left(p - \frac{1}{2} \times \frac{\partial p}{\partial x}\mathrm{d}x\right)\mathrm{d}y\mathrm{d}z$$

And the force exerted at c point on the opposite side is

$$P_c = p_c \mathrm{d}y\mathrm{d}z = \left(p + \frac{1}{2} \times \frac{\partial p}{\partial x}\mathrm{d}x\right)\mathrm{d}y\mathrm{d}z$$

In the x direction, resultant pressure is P_x

$$P_x = P_b - P_c = \left(p - \frac{1}{2} \times \frac{\partial p}{\partial x}\mathrm{d}x\right)\mathrm{d}y\mathrm{d}z - \left(p + \frac{1}{2} \times \frac{\partial p}{\partial x}\mathrm{d}x\right)\mathrm{d}y\mathrm{d}z = -\frac{\partial p}{\partial x}\mathrm{d}x\mathrm{d}y\mathrm{d}z$$

Similarly, in the y and z direction

$$P_y = -\frac{\partial p}{\partial y}\mathrm{d}x\mathrm{d}y\mathrm{d}z$$

$$P_z = -\frac{\partial p}{\partial z}\mathrm{d}x\mathrm{d}y\mathrm{d}z$$

(3) Force balance According to the condition of fluid equilibrium, the sum of the forces in any direction must be zero. Considering the x direction, there is $\sum \vec{F}_x = 0$.

$$f_x \rho \mathrm{d}x\mathrm{d}y\mathrm{d}z - \frac{\partial p}{\partial x}\mathrm{d}x\mathrm{d}y\mathrm{d}z = 0$$

Divide this equation by $\rho \mathrm{d}x\mathrm{d}y\mathrm{d}z$, and we can obtain

$$f_x - \frac{1}{\rho} \times \frac{\partial p}{\partial x} = 0$$

Similarly, in the y and z direction

$$\left.\begin{array}{l} f_x - \dfrac{1}{\rho} \times \dfrac{\partial p}{\partial x} = 0 \\ f_y - \dfrac{1}{\rho} \times \dfrac{\partial p}{\partial y} = 0 \\ f_z - \dfrac{1}{\rho} \times \dfrac{\partial p}{\partial z} = 0 \end{array}\right\} \quad (3\text{-}2)$$

This equation is called differential equation of fluid equilibrium. Euler firstly put forward the equation in 1755, so it is usually called Euler's equilibrium differential equation. When fluids are in equilibrium, the body force acting on the unit mass fluid equals to the resultant force of pressure.

3.2 流体平衡的微分方程式

3.2.1 欧拉平衡方程式

在图 3-3 中，流体中微元六面体的边长分别为 dx、dy 及 dz，体积为 dxdydz。流体密度为 ρ，在 a 点的压强为 $p(x, y, z)$。

作用在静止的流体微元上的力包括质量力和表面力。

(1) 质量力

质量力＝单位质量力×质量＝单位质量力× 密度 ×体积

x、y、z 方向的质量力分别为 F_x、F_y、F_z，表示如下

$$F_x = f_x \rho \mathrm{d}x \mathrm{d}y \mathrm{d}z$$

$$F_y = f_y \rho \mathrm{d}x \mathrm{d}y \mathrm{d}z$$

$$F_z = f_z \rho \mathrm{d}x \mathrm{d}y \mathrm{d}z$$

（2）表面力　既然流体是静止的，则微元的表面没有切应力，因此

$$\text{表面力} = \text{压强} \times \text{作用面的面积}$$

由于中心点的压强为 $p(x, y, z)$，与 x 轴正交的表面 b 点的力大约为

$$P_b = p_b \mathrm{d}y \mathrm{d}z = \left(p - \frac{1}{2} \times \frac{\partial p}{\partial x} \mathrm{d}x\right) \mathrm{d}y \mathrm{d}z$$

作用在对面的 c 点的力为

$$P_c = p_c \mathrm{d}y \mathrm{d}z = \left(p + \frac{1}{2} \times \frac{\partial p}{\partial x} \mathrm{d}x\right) \mathrm{d}y \mathrm{d}z$$

在 x 方向的压力的合力 P_x 为

$$P_x = P_b - P_c = \left(p - \frac{1}{2} \times \frac{\partial p}{\partial x} \mathrm{d}x\right) \mathrm{d}y \mathrm{d}z - \left(p + \frac{1}{2} \times \frac{\partial p}{\partial x} \mathrm{d}x\right) \mathrm{d}y \mathrm{d}z = -\frac{\partial p}{\partial x} \mathrm{d}x \mathrm{d}y \mathrm{d}z$$

同理，在 y、z 方向

$$P_y = -\frac{\partial p}{\partial y} \mathrm{d}x \mathrm{d}y \mathrm{d}z$$

$$P_z = -\frac{\partial p}{\partial z} \mathrm{d}x \mathrm{d}y \mathrm{d}z$$

（3）力的平衡　根据流体平衡的条件，在任何方向力的代数和应为 0。

在 x 方向，$\sum \vec{F}_x = 0$

$$f_x \rho \mathrm{d}x \mathrm{d}y \mathrm{d}z - \frac{\partial p}{\partial x} \mathrm{d}x \mathrm{d}y \mathrm{d}z = 0$$

上式除以 $\rho \mathrm{d}x \mathrm{d}y \mathrm{d}z$ 得

$$f_x - \frac{1}{\rho} \times \frac{\partial p}{\partial x} = 0$$

同理，在 y、z 方向，可得

$$\left. \begin{aligned} f_x - \frac{1}{\rho} \times \frac{\partial p}{\partial x} &= 0 \\ f_y - \frac{1}{\rho} \times \frac{\partial p}{\partial y} &= 0 \\ f_z - \frac{1}{\rho} \times \frac{\partial p}{\partial z} &= 0 \end{aligned} \right\} \quad (3\text{-}2)$$

这就是流体平衡微分方程，欧拉于 1755 年首先推导得到该方程，因此也被称为欧拉平衡微分方程。当流体平衡时，作用在单位质量流体上的质量力与压力的合力相平衡。

3.2.2　Potential Function of Body Forces

Multiply Equation (3-2) by $\mathrm{d}x$, $\mathrm{d}y$ and $\mathrm{d}z$ respectively and then summate the three equations, and we can obtain

$$f_x \mathrm{d}x + f_y \mathrm{d}y + f_z \mathrm{d}z - \frac{1}{\rho}\left(\frac{\partial p}{\partial x} \mathrm{d}x + \frac{\partial p}{\partial y} \mathrm{d}y + \frac{\partial p}{\partial z} \mathrm{d}z\right) = 0 \quad (3\text{-}3)$$

Since $p=p(x, y, z)$, then
$$dp = \rho(f_x dx + f_y dy + f_z dz) \tag{3-4}$$

Equation (3-4) is the general equation of Euler's equilibrium equation (differential equation of pressure).

For incompressible fluid, ρ is constant. From mathematical analysis theory we can know that the right-hand side in Equation (3-4) certainly is the whole differential of a certain coordinate function $W=W(x, y, z)$.

Let
$$\left. \begin{array}{l} f_x = -\dfrac{\partial W}{\partial x} \\ f_y = -\dfrac{\partial W}{\partial y} \\ f_z = -\dfrac{\partial W}{\partial z} \end{array} \right\} \tag{3-5}$$

From Equation (3-4) and Equation (3-5), we can obtain
$$\rho(f_x dx + f_y dy + f_z dz) = -\rho dW \tag{3-6}$$
Namely $$dp = -\rho dW$$

The coordinate function, $W=W(x, y, z)$, which satisfies Equation (3-5), is called a potential function of body force. The body force is called body force with potential. Fluids acted only by the body force with potential can keep equilibrium.

3.2.2 质量力的势函数

将式(3-2)分别乘以 dx、dy、dz 后相加，则有
$$f_x dx + f_y dy + f_z dz - \frac{1}{\rho}\left(\frac{\partial p}{\partial x}dx + \frac{\partial p}{\partial y}dy + \frac{\partial p}{\partial z}dz\right) = 0 \tag{3-3}$$

因 $p=p(x, y, z)$，则有
$$dp = \rho(f_x dx + f_y dy + f_z dz) \tag{3-4}$$

式(3-4)为欧拉平衡方程式的综合式(压强微分公式)。

对于不可压缩流体，ρ 为常数。根据数学分析理论可知，式(3-4) 右端也必是某一坐标函数 $W=W(x, y, z)$ 的全微分。

令
$$\left. \begin{array}{l} f_x = -\dfrac{\partial W}{\partial x} \\ f_y = -\dfrac{\partial W}{\partial y} \\ f_z = -\dfrac{\partial W}{\partial z} \end{array} \right\} \tag{3-5}$$

由式(3-4)、式(3-5) 可得
$$\rho(f_x dx + f_y dy + f_z dz) = -\rho dW \tag{3-6}$$
即 $$dp = -\rho dW$$

我们称满足式(3-5)的坐标函数 $W=W(x, y, z)$ 为质量力的势函数，而该质量力称为有势的质量力，流体只有在有势的质量力作用下才能平衡。

【**Sample Problem 3-1**】 Try to calculate the potential function of body force of static fluid in gravitational field.

Solution：Choose the coordinates system just as shown in Figure 3-4. Component forces

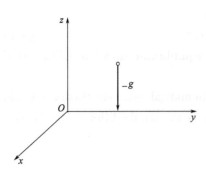

Figure 3-4 Body Force in Gravitational Field

图 3-4 在重力场中的质量分力

of unit mass are

$$f_x = f_y = 0, \quad f_z = -g$$

So

$$dW = \frac{\partial W}{\partial x}dx + \frac{\partial W}{\partial y}dy + \frac{\partial W}{\partial z}dz$$

$$= -(f_x dx + f_y dy + f_z dz) = g\,dz \tag{3-7}$$

Suppose the potential function is equal to zero on the coordinate plane ($z=0$). Namely, $W=0$ on the zero potential plane. Integrating, the potential function of body force of equilibrium fluid in the gravitational field is

$$W = gz \tag{3-8}$$

【例题 3-1】 试求重力场中平衡流体的质量力的势函数。

解： 取如图 3-4 所示的坐标系，则单位质量分力为

$$f_x = f_y = 0, \quad f_z = -g$$

于是

$$dW = \frac{\partial W}{\partial x}dx + \frac{\partial W}{\partial y}dy + \frac{\partial W}{\partial z}dz$$

$$= -(f_x dx + f_y dy + f_z dz) = g\,dz \tag{3-7}$$

设基准面 $z=0$ 处的势函数值为零，即零势面上 $W=0$，于是积分可得重力场中平衡流体的质量力的势函数为

$$W = gz \tag{3-8}$$

3.2.3 Equipressure Surface

The surface on which the pressures at all points are equivalent in a fluid is called equipressure surface. On equipressure surface the pressure p is constant, so

$$dp = 0$$

According to Equation (3-4), the differential equation of equipressure surface is

$$f_x dx + f_y dy + f_z dz = 0 \tag{3-9}$$

The characters of the equipressure surface are as follows.

(1) Equipressure surface is also an equipotential surface From Equation (3-6), we can know if $dp=0$, then $dW=0$. Thus

$$W = C \tag{3-10}$$

When the potential function of body force on the surface is constant, the surface is called equipotential surface. So equipressure surface is also an equipotential surface.

(2) Equipressure surface is vertical with unit body force vectors Write Equation (3-9) into the following vector form

$$\vec{f} \cdot d\vec{s} = 0 \tag{3-11}$$

where $d\vec{s}$—an arbitrary segment on equipressure surface. So equipressure surface is vertical with unit body force vectors.

3.2.3 等压面

流体中压强相等各点所组成的面叫作等压面。在等压面上，压强 p 是常数，则
$$dp = 0$$
由式(3-4)可得等压面的微分方程为
$$f_x dx + f_y dy + f_z dz = 0 \tag{3-9}$$
等压面的性质如下。

(1) 等压面也是等势面　由式(3-6)可见，$dp=0$ 时，$dW=0$，于是
$$W = C \tag{3-10}$$
质量力势函数等于常数的面叫作等势面，所以等压面也是等势面。

(2) 等压面与单位质量力矢量垂直　将式(3-9)写成矢量形式
$$\vec{f} \cdot d\vec{s} = 0 \tag{3-11}$$
式中，$d\vec{s}$ 是等压面上的任意线段，因而等压面与单位质量力矢量垂直。

3.3 Basic Equation of Fluid Statics

3.3.1 Basic Equation of Static Pressure

In Figure 3-5, Euler's equilibrium equation in a gravitational field can be written as
$$dp = -\rho dW = -\rho g dz \tag{3-12}$$

For a continuous, homogeneous and incompressible fluid the density is constant. So we can rewrite Equation (3-12) as
$$\int dp = \int -\rho g dz$$

Integrating
$$z + \frac{p}{\gamma} = C \tag{3-13}$$

where z——the vertical coordinate at any point in a static fluid;

p——the static pressure at any point in a static fluid.

Equation (3-13) is the basic equation of continuous, homogeneous and incompressible fluid in gravitational field.

According to the boundary conditions on free surface of equilibrium fluid, Equation (3-13) turns into
$$z + \frac{p}{\gamma} = z_0 + \frac{p_0}{\gamma} \qquad [3\text{-}14(a)]$$

Or
$$z_1 + \frac{p_1}{\gamma} = z_2 + \frac{p_2}{\gamma} \qquad [3\text{-}14(b)]$$

Moving the item
$$p = p_0 + \gamma(z_0 - z) = p_0 + \gamma h \qquad [3\text{-}15(a)]$$

Or

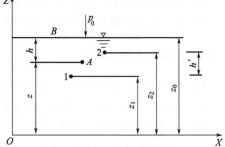

Figure 3-5　Equilibrium Fluid in Gravitational Field

图 3-5　重力场中的平衡流体

$$p_2 = p_1 + \gamma(z_1 - z_2) = p_1 + \gamma h' \qquad [3\text{-}15(b)]$$

According to Equation (3-15), the distributional rules of static pressure are as follows:

① The pressure p at any point in a static fluid is equal to the sum of surface pressure p_0 and the fluid weight from the point to the free surface on unit area $\rho g h$;

② In a static fluid, the pressure varies linearly with depth;

③ The equipressure surface acted on only by gravity in a static fluid is a horizontal plane.

3.3 流体静力学基本方程

3.3.1 静压强基本公式

如图 3-5 所示，重力场中的欧拉平衡方程式可以写为：

$$dp = -\rho dW = -\rho g\, dz \qquad (3\text{-}12)$$

对于连续、均质的不可压缩流体，其密度是常量。因而式(3-12) 变成

$$\int dp = \int -\rho g\, dz$$

积分，则得

$$z + \frac{p}{\gamma} = C \qquad (3\text{-}13)$$

式中，z 为平衡流体中任何一点的铅垂坐标；p 为平衡流体中任何一点的静压强。

式(3-13) 是重力场中连续、均质、不可压缩流体的静压强基本方程式。

根据平衡流体自由表面上的边界条件，式(3-13) 转化为

$$z + \frac{p}{\gamma} = z_0 + \frac{p_0}{\gamma} \qquad [3\text{-}14(a)]$$

或者

$$z_1 + \frac{p_1}{\gamma} = z_2 + \frac{p_2}{\gamma} \qquad [3\text{-}14(b)]$$

移项得

$$p = p_0 + \gamma(z_0 - z) = p_0 + \gamma h \qquad [3\text{-}15(a)]$$

或者

$$p_2 = p_1 + \gamma(z_1 - z_2) = p_1 + \gamma h' \qquad [3\text{-}15(b)]$$

根据式(3-15)，可得出静压分布规律：

① 静止流体中任一点的压强 p 等于表面压强 p_0 与从该点到流体自由表面的单位面积上的液体重量 $\rho g h$ 之和；

② 在静止流体中，压强随深度按线性规律变化；

③ 只受重力作用的静止流体中的等压面为水平面。

3.3.2 Units of Pressure

In engineering technology, pressure can be expressed with three units.

(1) Unit of stress It is expressed with force per unit area.

Units: Pa (Pascal) or N/m^2.

(2) Times of atmosphere pressure A standard atmospheric pressure is 1.013×10^5 Pa. It is equivalent to the pressure at the base of 760mm mercury column. An engineering atmospheric pressure is 0.98×10^5 Pa. It is equivalent to the pressure at the base of 10m water column.

Units: Times.

(3) The height of liquid column   The relation between the height of liquid column and pressure is $p=\gamma h$ or $h=p/\gamma$.

Units: m of H_2O or mm of Hg.

3.3.2 压强的计量单位

在工程技术上，常用如下三种方法表示压强的单位。

(1) 应力单位   采用单位面积上承受的力来表示，单位为 Pa 或 N/m^2。

(2) 用大气压强的倍数表示   一个标准大气压是 $1.013\times10^5 Pa$，相当于760mm 汞柱所产生的压强。一个工程大气压是 $0.98\times10^5 Pa$，相当于10m 水柱所产生的压强。单位为倍数。

(3) 液柱高单位   液柱高与压强的关系为 $p=\gamma h$ 或 $h=p/\gamma$，单位为 m 水柱或 mm 汞柱。

3.3.3 Expression Methods of Pressure

Pressure can be expressed with reference to a datum. The usual datum is absolute zero or local atmospheric pressure. The relationships of different pressures are shown in Figure 3-6.

(1) Absolute pressure   If a pressure is expressed as a difference between its value and absolute zero, it is called an absolute pressure p_{abs}.

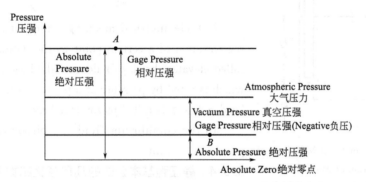

Figure 3-6   Expression Methods of Pressure

图 3-6   压强的表示方法

(2) Gage pressure   If a pressure is expressed as a difference between its value and local atmospheric pressure p_a, it is called a gage pressure p. The relation between absolute pressure p_{abs} and gage pressure p is

$$p=p_{abs}-p_a \tag{3-16}$$

(3) Vacuum pressure   If the pressure is below the local atmospheric pressure, the vacuum value p_v is expressed as a difference between local atmospheric pressure p_a and absolute pressure p_{abs}. The relation is

$$p_v=p_a-p_{abs} \tag{3-17}$$

The height of liquid column for vacuum value p_v is called degree of vacuum h_v.

$$h_v=\frac{p_v}{\gamma} \tag{3-18}$$

3.3.3 压强的表示方法

压强可以参照某些基准点来表示，通常以绝对零点或当地大气压作为基准点。图 3-6 表

(1) 绝对压强　以绝对零点为基准起算的压强，叫作绝对压强 p_{abs}。

(2) 相对压强　相对压强就是以当地大气压强 p_a 为零起算的压强，又称表压强 p。绝对压强与相对压强的关系为

$$p = p_{abs} - p_a \quad (3-16)$$

(3) 真空压强　假如某点压强小于当地大气压强，呈现真空状态，当地大气压与该点绝对压强的差值 $p_a - p_{abs}$ 称为真空值 p_v，即

$$p_v = p_a - p_{abs} \quad (3-17)$$

真空值为 p_v 的液柱高度称为真空度 h_v。

$$h_v = \frac{p_v}{\gamma} \quad (3-18)$$

3.3.4 The Geometrical and Physical Meanings of Basic Equation of Static Pressure

The manometric tube is shown in Figure 3-7, and plane 0-0 is the datum plane. According to Equation (3-14), we can know

$$z_1 + \frac{p_1}{\gamma} = z_2 + \frac{p_2}{\gamma}$$

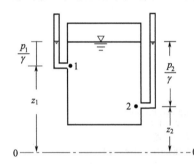

Figure 3-7 Geometrical Meanings of the Basic Equation of Static Pressure

图 3-7　静压强基本公式的几何意义

(1) Geometrical meaning　z denotes the height from the elevation of a certain point to the datum plane, so it is called elevation head. p/γ denotes the height of liquid column which is acted by pressure at a point, so it is called pressure head. $z + p/\gamma$ is called piezometer head.

In the equilibrium fluid the piezometer head at any point is constant.

3.3.4 静压强基本公式的几何意义和物理意义

如图 3-7 所示的测压管，令基准面为 0—0，由式 (3-14) 可得

$$z_1 + \frac{p_1}{\gamma} = z_2 + \frac{p_2}{\gamma}$$

(1) 几何意义　z 表示某点位置到基准面的高度，称为位置水头；p/γ 表示某点压强作用下的液柱高度，称为压强水头；$z + p/\gamma$ 称为测压管水头。

平衡流体中，各点的测压管水头是一常数。

(2) Physical meaning　In a physical view, z denotes the potential energy per unit weight fluid. $\frac{p}{\gamma}$ denotes the pressure energy per unit weight fluid. $z + \frac{p}{\gamma}$ denotes the whole potential energy per unit weight fluid.

In the equilibrium fluid the whole potential energy is constant.

(2) 物理意义　从物理学角度看，z 表示单位重量流体的位置势能，简称单位位能；$\frac{p}{\gamma}$ 表示单位重量流体的压强势能，简称单位压能；$z + \frac{p}{\gamma}$ 表示单位重量流体的总势能，简称单

位势能。

平衡流体中各点的单位势能是一常数。

3.3.5 Diagram of Static Pressure Distribution

As shown in Figure 3-8 (a), a liquid is contained in the tank with a vertical wall. The gage pressure is zero at the point A on the free surface. At the depth of H, $p=\gamma H$. The relationship between p and H is linear and can be represented by the triangle. The lines drawn in the diagram are perpendicular to the acting surface. The length of the lines represents the magnitude of the pressure and the arrow represents the direction of the pressure at the point. Extend from $p=0$ at A on the free surface to $p=\gamma H$ at depth H, and a straight line will appear. This is called diagram of static pressure distribution. The diagrams of pressure distribution on other surfaces are shown in Figure 3-8(b)~(d).

3.3.5 静压分布图

在图 3-8(a) 中，具有垂直墙壁的容器盛有液体。自由表面的 A 点相对压强是 0，在深度 H 处，$p=\gamma H$。p 和 H 是线性关系，且可以用三角形来表示。图中所画直线垂直于作用面，线段的长度代表某点压强的大小，箭头代表压强的方向。从自由表面的 $p=0$ 点到深度 H 处的 $p=\gamma H$ 可以连成一条直线。这就叫静压分布图。其他作用面的静压分布图如图 3-8 (b)~(d) 所示。

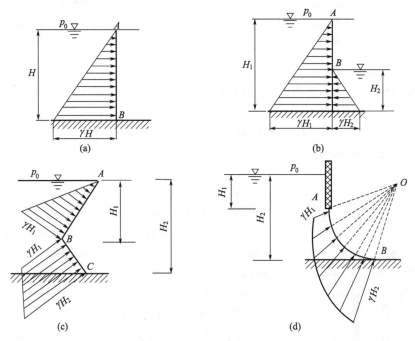

Figure 3-8 Diagrams of Static Pressure Distribution
图 3-8 静压分布图

3.4 Measurement of Static Pressure

The measure instruments of static pressure of fluid are metal, electrical logging and liquid column. The measurement precision of liquid column instrument is high and the range is small. It is suitable for the experiment place with low pressure. The typic liquid column instruments are as follows.

(1) Manometric tube　As shown in Figure 3-9, the gage pressure in water surface can be measured: $p_1 = \rho g h$.

(2) U-tube manometer　As shown in Figure 3-10, the U-tube manometer can be used to measure the pressure of either liquids or gases. The bottom of the U-tube manometer is filled with a liquid M which is of specific weight γ_m and is immiscible with the fluid W, liquid or gas of specific weight γ. If E is a point of interface in the left-hand and F is a point at the same level in the right-hand tube, Pressure p_E at E = Pressure p_F at F.

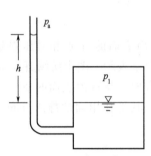

Figure 3-9　Manometric Tube
图 3-9　测压管

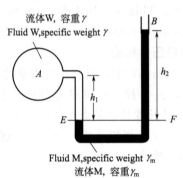

Figure 3-10　U-tube Manometer
图 3-10　U 形管测压计

For the left-hand tube
p_E = Pressure p_A at A + Pressure due to depth h_1 of fluid W = $p_A + \gamma h_1$
For the right-hand tube
p_F = Pressure p_B at B + Pressure due to depth h_2 of liquid M = $p_B + \gamma_m h_2$
$p_B = 0$ (Gage pressure), so $p_F = 0 + \gamma_m h_2$
Since $p_E = p_F$,

$$p_A + \gamma h_1 = 0 + \gamma_m h_2,$$
$$p_A = \gamma_m h_2 - \gamma h_1$$

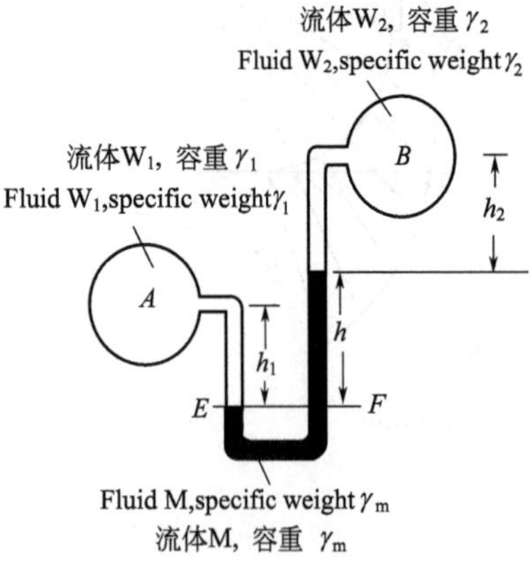

(3) U-tube differential manometer　As shown in Figure 3-11, the U-tube differential manometer can be used to measure the differential pressure between A point and B point.

For the left-hand tube
p_E = Pressure p_A at A + Pressure due to depth h_1 of fluid W_1 = $p_A + \gamma_1 h_1$
For the right-hand tube
p_F = Pressure p_B at B + Pressure due to depth h_2 of fluid W_2 + Pressure due to depth h of fluid M = $p_B + \gamma_2 h_2 + \gamma_m h$
Since $p_E = p_F$

Figure 3-11　U-tube Differential Manometer
图 3-11　U 形管压差计

$$p_A+\gamma_1h_1=p_B+\gamma_2h_2+\gamma_m h$$
$$p_A-p_B=\gamma_2h_2+\gamma_m h-\gamma_1h_1$$

(4) Micro-manometer　The instrument being used to measure the less pressure or pressure difference is called micro-manometer. One of this kind of instruments is shown in Figure 3-12. The inclined micro-manometer consists of a glass tube (the area of cross section is A_1) whose obliquity α can be adjusted and a small container with liquid (the area of cross section is A_2). If the entrance pressure p_1 of the inclined tube equals to the entrance pressure p_2 of the container, the liquid surfaces in the container and in the inclined tube are even. When p_1 doesn't equal to p_2 such as $p_1 < p_2$, the liquid surface of the inclined tube will ascend h and the liquid surface of the container will descend Δh. According to Equation(3-15)

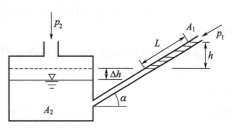

Figure 3-12　Inclined Micro-Manometer
图 3-12　倾斜式微压计

$$p_2=p_1+\rho g(h+\Delta h) \tag{3-19}$$

Because the descending volume in the container equals to the ascending volume in the inclined tube, namely

$$\Delta h=\frac{A_1}{A_2}L \tag{3-20}$$

and
$$h=L\sin\alpha \tag{3-21}$$

According to Equation (3-19) to Equation (3-21), We can obtain

$$p_2-p_1=\rho g\left(\sin\alpha+\frac{A_1}{A_2}\right)L \tag{3-22}$$

3.4　流体静压强的测量

流体静压强的测量仪器有金属式、电测式、液柱式等。液柱式仪表测量精度高，量程小，适用于低压实验场所。常用的液柱式仪表有如下几种形式。

(1) 测压管　如图 3-9 所示，可测得水面的相对压强：$p_1=\rho gh$。

(2) U 形管测压计　如图 3-10 所示，U 形管测压计可以用于测量液体或气体的压强，U 形管的底部装有容重为 γ_m 的液体 M，与互不相容的液体或气体 W 相连，其容重为 γ。假如 E、F 两点分别是同一水平面上左右两肢上的点，则

$$E\text{ 点压强 } p_E = F \text{ 点压强 } p_F$$

对于左肢
$$p_E = A \text{ 点压强 } p_A + \text{流体 W 高度 } h_1 \text{ 产生的压强} = p_A+\gamma h_1$$

对于右肢
$$p_F = B \text{ 点压强 } p_B + \text{流体 M 高度 } h_2 \text{ 产生的压强} = p_B+\gamma_m h_2$$
$$p_B=0 \text{（相对压强）}$$

因此
$$p_F=0+\gamma_m h_2$$
由于
$$p_E=p_F$$
可得
$$p_A+\gamma h_1=0+\gamma_m h_2$$
$$p_A=\gamma_m h_2-\gamma h_1$$

(3) U 形管压差计 如图 3-11 所示，U 形管压差计可以用来测量 A、B 两点的压强差。
对于左肢
$$p_E = A \text{ 点的压强 } p_A + \text{流体 } W_1 \text{ 高度 } h_1 \text{ 产生的压强} = p_A + \gamma_1 h_1$$
对于右肢
$$p_F = B \text{ 点的压强 } p_B + \text{流体 } W_2 \text{ 高度 } h_2 \text{ 产生的压强} + \text{流体 M 高度 } h \text{ 产生的压强}$$
$$= p_B + \gamma_2 h_2 + \gamma_m h$$
由于
$$p_E = p_F$$
因此
$$p_A + \gamma_1 h_1 = p_B + \gamma_2 h_2 + \gamma_m h$$
$$p_A - p_B = \gamma_2 h_2 + \gamma_m h - \gamma_1 h_1$$

(4) 微压计 测量较小压强或压强差的仪器叫作微压计，图 3-12 就是其中一种。倾斜式微压计由一根倾角 α 可调的玻璃管（横截面面积为 A_1）和一个盛液体的小容器（横截面面积为 A_2）组成。如果斜管入口压强 p_1 和容器入口压强 p_2 相等，则容器内液面与斜管中的液面齐平；当 p_1 和 p_2 不相等时，例如 $p_1 < p_2$，则斜管中液面将上升 h，容器内液面下降 Δh。因此，由式 (3-15) 可得

$$p_2 = p_1 + \rho g (h + \Delta h) \tag{3-19}$$

由于容器内液面下降的体积与斜管中液面上升的体积相等，即有

$$\Delta h = \frac{A_1}{A_2} L \tag{3-20}$$

又

$$h = L \sin \alpha \tag{3-21}$$

由式 (3-19) ~ 式 (3-21) 得

$$p_2 - p_1 = \rho g \left(\sin \alpha + \frac{A_1}{A_2} \right) L \tag{3-22}$$

3.5 Relative Equilibrium of Liquid

A container with liquid has relative motion to the fixed coordinate system on earth, but there is no relative motion between the liquid particles. This motion state is called relative equilibrium. Besides the equilibrium problems of fluid in gravitational field, in engineering there are two kinds of circumstances: a linear motional container with liquid at a uniform acceleration and a rotating container with liquid at constant angular velocity.

3.5 液体的相对平衡

若盛液体的容器对地面上的固定坐标系有相对运动，但液体质点彼此之间却没有相对运动，这种运动状态称为相对平衡。除了重力场的流体平衡问题外，工程上常见的有如下两种：容器做匀加速直线运动和容器做等角速回转运动。

3.5.1 Container's Linear Motion at a Uniform Acceleration

As shown in Figure 3-13, a container with liquid does linear motion downwards at a uniform acceleration along the inclined plane that makes an angle of α with the horizontal.

According to the d'Alembert paradox, the body forces on the equilibrium fluid particles include the fictitious inertial force contrary to the direction of acceleration and gravity.

According to Figure 3-13, the component forces per unit mass are

$$\left.\begin{aligned} f_x &= 0 \\ f_y &= a\cos\alpha \\ f_z &= a\sin\alpha - g \end{aligned}\right\} \quad (3\text{-}23)$$

where f_x, f_y and f_z—the component forces per unit mass in the x, y and z direction, respectively;

a—the acceleration of the container in the motion direction.

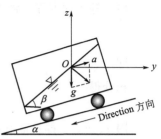

Figure 3-13 Linear Motion at a Uniform Acceleration

图 3-13 匀加速直线运动

(1) Equation of equipressure surface Substituting Equation (3-23) into Equation (3-9) yields

$$a\cos\alpha \, \mathrm{d}y + (a\sin\alpha - g)\mathrm{d}z = 0$$

Namely

$$\frac{\mathrm{d}z}{\mathrm{d}y} = \frac{a\cos\alpha}{g - a\sin\alpha} = \tan\beta \quad (3\text{-}24)$$

Integrating

$$a\cos\alpha \, y + (a\sin\alpha - g)z = c \quad (3\text{-}25)$$

Since α, g and a are all constants, β is also a constant.

The equipressure surfaces (including free surface) are a group of parallel surfaces which make an angle β with the horizontal. Moreover, these surfaces should keep vertical with the unit body force f_m.

(2) Distribution rules of static pressure Substituting Equation (3-25) into Equation (3-4) yields

$$\mathrm{d}p = \rho[a\cos\alpha \, \mathrm{d}y + (a\sin\alpha - g)\mathrm{d}z]$$

Integrating

$$p = \rho[ay\cos\alpha + z(a\sin\alpha - g)] + C$$

According to the boundary conditions $y=0$, $z=0$, $p=p_0$, we obtain

$$p = p_0 + \rho[ay\cos\alpha + z(a\sin\alpha - g)] \quad (3\text{-}26)$$

As shown in Figure 3-14, when $\alpha=0$ or $\pi/2$, we can get the distributional rule of static pressure during the container does horizontal or vertical motion at a uniform acceleration.

3.5.1 容器做匀加速直线运动

如图 3-13 所示，盛有液体的容器沿着与水平基面成 α 角的斜面向下以匀加速度 a 做直线运动。

根据达朗贝尔佯谬，相对平衡流体质点上的质量力为与加速度方向相反的虚构惯性力与重力。由图 3-13 可得单位质量分力为

$$\left.\begin{aligned} f_x &= 0 \\ f_y &= a\cos\alpha \\ f_z &= a\sin\alpha - g \end{aligned}\right\} \quad (3\text{-}23)$$

式中，f_x、f_y、f_z 分别表示在 x、y、z 方向的单位质量力；a 为容器在运动方向的加速度。

(1) 等压面方程 将式(3-23)代入式(3-9)可得

$$a\cos\alpha\,dy + (a\sin\alpha - g)\,dz = 0$$

即

$$\frac{dz}{dy} = \frac{a\cos\alpha}{g - a\sin\alpha} = \tan\beta \tag{3-24}$$

积分得

$$a\cos\alpha\, y + (a\sin\alpha - g)z = c \tag{3-25}$$

因 α、g、a 都是常数，故 β 是一定值。等压面（包括自由表面）是与水平基面成倾角 β 的一族平行平面，这族平面应与单位质量力 f_m 相垂直。

(2) 静压强分布规律 将式(3-25)代入式(3-4)中即得

$$dp = \rho[a\cos\alpha\,dy + (a\sin\alpha - g)dz]$$

作不定积分得

$$p = \rho[ay\cos\alpha + z(a\sin\alpha - g)] + C$$

根据边界条件 $y=0$、$z=0$、$p=p_0$ 可得

$$p = p_0 + \rho[ay\cos\alpha + z(a\sin\alpha - g)] \tag{3-26}$$

当 $\alpha = 0$ 或 $\pi/2$ 时，即可得出容器水平或竖向匀加速直线运动时的静压分布规律，如图 3-14 所示。

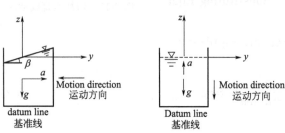

Figure 3-14 The Horizontal and Vertical Motion at a Uniform Acceleration

图 3-14 水平和竖向的匀加速直线运动

【Sample Problem 3-2】 As shown in Figure 3-15, a watering cart is moving along with a uniform linear acceleration, $a = 0.98 \text{m/s}^2$. Try to calculate the inclination angle α between the free surface and the horizontal. Assuming the water depth at B point at rest $h = 1.0\text{m}$, and $x_B = -1.5\text{m}$, what is the static pressure of this point in the moving watering cart?

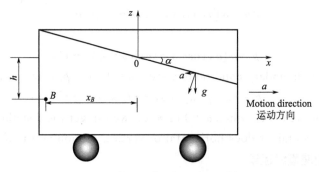

Figure 3-15 Figure for Sample Problem 3-2

图 3-15 例题 3-2 图

Solution: Unit body forces of gravitation are

$$f_{x_1} = f_{y_1} = 0, \quad f_{z_1} = -g$$

Unit body forces of inertial force are

$$f_{x_2} = -a, \quad f_{y_2} = f_{z_2} = 0$$

So the resultant unit body forces are

$$f_x = f_{x_1} + f_{x_2} = -a$$
$$f_y = f_{y_1} + f_{y_2} = 0$$

$$f_z = f_{z_1} + f_{z_2} = -g$$

Substituting into Equation(3-4), we obtain

$$dp = \rho(-a\,dx - g\,dz)$$

Integrating

$$p = -\rho(ax + gz) + C$$

For $x = z = 0$, $p = p_0$, there is $C = p_0$, namely

$$p = p_0 - \gamma\left(\frac{a}{g}x + z\right)$$

The gage pressure at B point is

$$p = -\gamma\left(\frac{a}{g}x_B + z_B\right) = -9800\left[\frac{0.98}{9.80} \times (-1.5) + (-1.0)\right] = 11270\,\text{N/m}^2 = 11.27\,\text{kPa}$$

The formula for free surface is

$$ax + gz = 0$$

Thus

$$\tan\alpha = -\frac{z}{x} = \frac{a}{g} = \frac{0.98}{9.80} = 0.10$$

So

$$\alpha = 5°45'$$

【例题 3-2】 一辆洒水车以等加速度 $a = 0.98\,\text{m/s}^2$ 向前行驶，如图 3-15 所示，求水车内自由表面与水平面间的夹角 α；若 B 点在运动前位于水面下深为 $h = 1.0\,\text{m}$ 处，距 z 轴 $x_B = -1.5\,\text{m}$，求洒水车加速运动后该点的静水压强。

解：重力的单位质量力为 $f_{x_1} = f_{y_1} = 0$，$f_{z_1} = -g$。惯性力的单位质量力为 $f_{x_2} = -a$，$f_{y_2} = f_{z_2} = 0$。总的单位质量力为

$$f_x = f_{x_1} + f_{x_2} = -a$$
$$f_y = f_{y_1} + f_{y_2} = 0$$
$$f_z = f_{z_1} + f_{z_2} = -g$$

代入式(3-4) 得

$$dp = \rho(-a\,dx - g\,dz)$$

积分得

$$p = -\rho(ax + gz) + C$$

当 $x = z = 0$ 时，$p = p_0$，得 $C = p_0$，代入上式得

$$p = p_0 - \gamma\left(\frac{a}{g}x + z\right)$$

B 点的相对压强为

$$p = -\gamma\left(\frac{a}{g}x_B + z_B\right) = -9800\left[\frac{0.98}{9.80} \times (-1.5) + (-1.0)\right] = 11270\,\text{N/m}^2 = 11.27\,\text{kPa}$$

而自由液面方程为

$$ax + gz = 0$$

即

$$\tan\alpha = -\frac{z}{x} = \frac{a}{g} = \frac{0.98}{9.80} = 0.10$$

故得

$$\alpha = 5°45'$$

3.5.2 Container's Rotation at a Constant Angular Speed

As shown in Figure 3-16, the container with liquid is rotating around the z axis. When

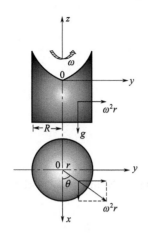

Figure 3-16 Container's Rotation at a Constant Angular Speed
图 3-16 容器做等角速回转运动

the rotation motion has been stable, the free surface of liquid will be formed just as shown in Figure 3-16, and there are no relative motions between liquid mass particles. With the similar analytical method of the linear motional container at uniform acceleration, the components of unit body force are

$$\left.\begin{array}{l} f_x = \omega^2 r\cos\theta = \omega^2 x \\ f_y = \omega^2 r\sin\theta = \omega^2 y \\ f_z = -g \end{array}\right\} \quad (3\text{-}27)$$

(1) Equation of equipressure surface Substitute Equation (3-27) into Equation (3-9) yields

$$\omega^2 x\,dx + \omega^2 y\,dy - g\,dz = 0$$

Integrating

$$\frac{\omega^2 x^2}{2} + \frac{\omega^2 y^2}{2} - gz = c$$

Namely

$$\frac{\omega^2 r^2}{2} - gz = c \quad (3\text{-}28)$$

The equipressure surfaces are a group of rotational paraboloids about the z axis.

(2) Distribution rules of static pressure Substitute Equation (3-27) into Equation (3-4) yields

$$dp = \rho(\omega^2 x\,dx + \omega^2 y\,dy - g\,dz)$$

Integrate, and then we obtain

$$p = \rho\left(\frac{\omega^2 x^2}{2} + \frac{\omega^2 y^2}{2} - zg\right) + c = \rho\left(\frac{\omega^2 r^2}{2} - zg\right) + c \quad (3\text{-}29)$$

The integral constant in Equation (3-29) can be determined by the following three circumstances.

① For the case of a closed container, the pressure on liquid surface is p_0 (as shown in Figure 3-17).

Substitute the boundary conditions $r=0$, $z=0$, $p=p_0$ into Equation (3-29), and the result is

$$p = p_0 + \rho g\left(\frac{\omega^2 r^2}{2g} - z\right) \quad (3\text{-}30)$$

② The container is filled with liquid and the center of coping contacts atmosphere (as shown in Figure 3-18).

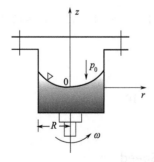

Figure 3-17 A Closed Container
图 3-17 密封容器

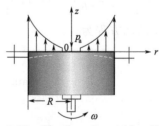

Figure 3-18 The Container with a Hatch in the Center of Coping
图 3-18 顶盖中心开口容器

Substitute the boundary conditions $r=0$, $z=0$, $p=p_a$ into Equation (3-29) and we obtain

$$p=p_a+\rho g\left(\frac{\omega^2 r^2}{2g}-z\right) \tag{3-31}$$

③ The container is filled with liquid and the borders of coping contacts atmosphere (as shown in Figure 3-19).

Substitute the boundary conditions $r=R$, $z=0$, $p=p_a$ into Equation (3-29) and we obtain

$$p=p_a-\rho g\left[\frac{\omega^2}{2g}(R^2-r^2)+z\right] \tag{3-32}$$

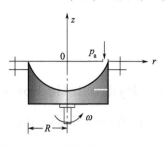

Figure 3-19 The Container with a Hatch on the Border of Coping

图 3-19 顶盖边缘开口容器

3.5.2 容器做等角速回转运动

如图 3-16 所示，盛有液体的容器绕铅直轴 z 做回转运动，待运动稳定后，液体形成如图所示的自由表面，质点之间不再有相对运动。与容器做匀加速直线运动的分析相同，单位质量分力为：

$$\left.\begin{aligned} f_x &= \omega^2 r\cos\theta = \omega^2 x \\ f_y &= \omega^2 r\sin\theta = \omega^2 y \\ f_z &= -g \end{aligned}\right\} \tag{3-27}$$

（1）等压面方程 将式(3-27)代入式(3-9)中，得

$$\omega^2 x\,\mathrm{d}x + \omega^2 y\,\mathrm{d}y - g\,\mathrm{d}z = 0$$

做不定积分得

$$\frac{\omega^2 x^2}{2}+\frac{\omega^2 y^2}{2}-gz=c$$

即

$$\frac{\omega^2 r^2}{2}-gz=c \tag{3-28}$$

等压面是一族绕 z 轴的旋转抛物面。

（2）静压强分布规律 将式(3-27)代入式(3-4)中得

$$\mathrm{d}p=\rho(\omega^2 x\,\mathrm{d}x+\omega^2 y\,\mathrm{d}y-g\,\mathrm{d}z)$$

做不定积分得

$$p=\rho\left(\frac{\omega^2 x^2}{2}+\frac{\omega^2 y^2}{2}-zg\right)+c=\rho\left(\frac{\omega^2 r^2}{2}-zg\right)+c \tag{3-29}$$

式中的积分常数可以根据如下三种情况来确定。

① 密封容器，液面上的压强为 p_0（图 3-17）。

将边界条件 $r=0$、$z=0$、$p=p_0$ 代回式(3-29)得

$$p=p_0+\rho g\left(\frac{\omega^2 r^2}{2g}-z\right) \tag{3-30}$$

② 容器盛满液体，顶盖中心接触大气（图 3-18）。

将边界条件 $r=0$、$z=0$、$p=p_a$ 代回式(3-29)得

$$p=p_a+\rho g\left(\frac{\omega^2 r^2}{2g}-z\right) \tag{3-31}$$

③ 容器盛满液体，顶盖边缘接触大气（图 3-19）。

将边界条件 $r=R$、$z=0$、$p=p_a$ 代回式(3-29)得

$$p = p_a - \rho g \left[\frac{\omega^2}{2g}(R^2 - r^2) + z\right] \tag{3-32}$$

3.6　Hydrostatic Forces on Plane Surfaces

The force acting on the wall of static fluid is called hydrostatic force.

3.6　作用在平面上的静水总压力

静止流体作用在壁面上的力称为静水总压力。

3.6.1　Graphical Method

(1) The magnitude of total pressure force　ABE is the pressure diagram for the vertical wall of the tank containing a liquid in Figure 3-20 (a), and $ABEA'B'E'$ is the pressure distribution volume in Figure 3-20 (b). If the force acting on dA is dF, then

$$dF = p\,dA$$

$$F = \int dF = \int_{ABB'A'} p\,dA = \int_{ABB'A'} \gamma h\,dA = \text{Volume of } ABEA'B'E' \tag{3-33}$$

Thus, the resultant force $F =$ (Area of ABE) $\times$ (Width b), namely

$$F = \frac{1}{2}\gamma H^2 b \tag{3-34}$$

where H is the vertical depth of liquid.

Because the vertical depth to the centroid H_C is half of H and the area is Hb, then

$$F = \gamma H_C A = p_C A \tag{3-35}$$

where p_C is the pressure of the centroid.

(2) The direction of total pressure force　The resultant force F on the immersed surface is normal to the plane. The direction is horizontal and rightword.

(3) Acting point of total pressure force　The resultant force F will act through the centroid of the pressure diagram, which is at a depth of $\frac{2}{3}H$ from A, The acting point is D.

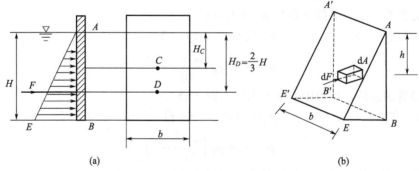

Figure 3-20　Calculation of Hydrostatic Force Using Graphical Method
图 3-20　图解法求平面静水总压力

$$H_D = \frac{2}{3}H$$

3.6.1 图解法

(1) 总压力的大小　图 3-20(a) 中，ABE 是静压分布图。图 3-20(b) 中，$ABEA'B'E'$ 是静压分布体，dF 为作用在 dA 上的力。

$$dF = p\,dA$$

$$F = \int dF = \int_{ABB'A'} p\,dA = \int_{ABB'A'} \gamma h\,dA = ABEA'B'E' \text{静压分布体的体积} \quad (3\text{-}33)$$

因此，合力 $F = (ABE\text{ 的面积}) \times (\text{宽度 } b)$，即

$$F = \frac{1}{2}\gamma H^2 b \quad (3\text{-}34)$$

式中，H 为液体的垂直深度。

因为作用面形心点 H_C 是 H 的一半，又作用面的面积是 Hb，则

$$F = \gamma H_C A = p_C A \quad (3\text{-}35)$$

式中，p_C 为形心点压强。

(2) 总压力的方向　方向垂直并指向作用面，水平向右。

(3) 总压力的作用点　合力作用点必通过静压分布图的形心处，即在 A 点液面下 $2H/3$ 处，作用点为 D 点。

$$H_D = 2H/3$$

3.6.2 Analytic Method

(1) The magnitude of total pressure force　As shown in Figure 3-21, plane A makes an angle α with the liquid surface.

If the infinitesimal area is equal to dA, then the hydrostatic force on the infinitesimal area is

$$dF = p\,dA = \rho g h\,dA = \rho g \sin\alpha\, y\,dA \quad (3\text{-}36)$$

Summate to the equilibrium system of forces and then the hydrostatic force on plane A is

$$F = \int_A dF = \rho g \sin\alpha \int_A y\,dA \quad (3\text{-}37)$$

In Equation (3-37), $\int_A y\,dA$ denotes the area moment of area A about the axis ox, and it is equal to the product of area A and the coordinate of the centroid y_C. If p_C is the static pressure at the centroid C, then

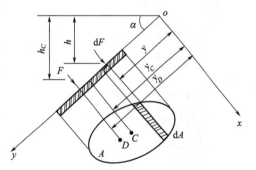

Figure 3-21　Calculation of Hydrostatic Force Using Analytic Method

图 3-21　解析法求平面静水总压力

$$F = \rho g \sin\alpha\, y_C A = p_C A \quad (3\text{-}38)$$

The resultant force $F = $ (the pressure at its centroid p_C) $\times$ (Area of plane A)

Force on the plane with arbitrary shape is equal to the product of the area of the plane and the pressure at its centroid.

(2) The direction of total pressure force　According to the characters of static pressure, it is certain that the resultant force F points to the acting surface vertically.

(3) Acting point of total pressure force   The acting point of total pressure force is called the center of pressure, marked as D. The coordinate in direction y of press center D is

$$y_D = \frac{\int_A y\,dF}{F} = \frac{\rho g \sin\alpha \int_A y^2\,dA}{\rho g h_C A} = \frac{\rho g \sin\alpha I_x}{\rho g \sin\alpha y_C A} = \frac{I_x}{y_C A} \tag{3-39}$$

where $\int_A y^2\,dA = I_x$, and it is the moment of inertia of area A of plane about the axis ox.

According to the parallel axis theorem of the moment of inertia:

$$I_x = y_C^2 A + I_C \tag{3-40}$$

Substitute it into Equation (3-39) yields

$$y_D = \frac{I_C}{y_C A} + y_C \tag{3-41}$$

For
$$\frac{I_C}{y_C A} > 0$$

so $y_D > y_C$. The center of pressure D is below the centroid C, and the distance between them is equal to $\dfrac{I_C}{y_C A}$.

Moment of inertia, centroid and area for plane figure is shown in Table 3-1.

Table 3-1   Moment of Inertia, Centroid and Area for Plane Figure

表 3-1   平面图形的惯性矩、形心及面积

Plane 平面	Moment of inertia 惯性矩 J_C	Centroid 形心 h_C	Area 面积 A
rectangle	$\dfrac{1}{12}bh^3$	$\dfrac{1}{2}h$	bh
triangle	$\dfrac{1}{36}bh^3$	$\dfrac{2}{3}h$	$\dfrac{1}{2}bh$
circle	$\dfrac{1}{4}\pi r^4$	r	πr^2
trapezoid	$\dfrac{h^3(a^2+4ab+b^2)}{36(a+b)}$	$\dfrac{h(a+2b)}{3(a+b)}$	$\dfrac{h(a+b)}{2}$

3.6.2 解析法

（1）总压力的大小　如图 3-21 所示，平面 A 与液面倾斜成 α 角，取微元面积 dA，则微元面积上的流体静压力大小为

$$dF = p\,dA = \rho g h\,dA = \rho g \sin\alpha\, y\,dA \tag{3-36}$$

对平衡力系求和，则可得平面 A 上的总压力为

$$F = \int_A dF = \rho g \sin\alpha \int_A y\,dA \tag{3-37}$$

式中，$\int_A y\,dA$ 代表面积 A 对 ox 轴的面积矩，它等于面积 A 与其形心坐标 y_C 的乘积。以 p_C 代表形心 C 处的静水压强，则

$$F = \rho g \sin\alpha\, y_C A = p_C A \tag{3-38}$$

总压力大小 $F =$ 作用面形心点压强 $p_C \times$ 作用面面积 A

作用在任意形状平面上的总压力大小等于该平面的面积与其形心处压力的乘积。

（2）总压力的方向　根据静压力的特性，总压力 F 必然垂直地指向这个作用面。

（3）总压力的作用点　总压力的作用点称为压力中心，记作 D 点。压力中心 D 在 y 方向上的坐标

$$y_D = \frac{\int_A y\,dF}{F} = \frac{\rho g \sin\alpha \int_A y^2\,dA}{\rho g h_C A} = \frac{\rho g \sin\alpha I_x}{\rho g \sin\alpha y_C A} = \frac{I_x}{y_C A} \tag{3-39}$$

式中，$\int_A y^2\,dA = I_x$ 是平面面积 A 对 ox 轴的惯性矩。

根据惯性矩的平行移轴定理

$$I_x = y_C^2 A + I_C \tag{3-40}$$

将式（3-40）代回式（3-39），则

$$y_D = \frac{I_C}{y_C A} + y_C \tag{3-41}$$

因为 $\dfrac{I_C}{y_C A} > 0$，所以 $y_D > y_C$，即压力中心 D 恒在平面形心 C 的下方，其间距为 $\dfrac{I_C}{y_C A}$。

平面图形的惯性矩、形心及面积见表 3-1。

【**Sample Problem 3-3**】 As shown in Figure 3-22 (a), there is a rectangle gate where length $L = 3$m, width $b = 2$m, and inclination $\alpha = 30°$. Please calculate the total pressure acting on the gate using graphical method and analytic method, respectively.

Solution：

Method 1　Graphical method is shown in Figure 3-22 (b).

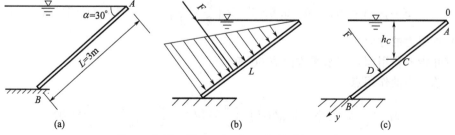

Figure 3-22　Figure for Sample Problem 3-3

图 3-22　例题 3-3 图

(1) The magnitude of total pressure force

$$F = \frac{1}{2} \times \left(\gamma \times \frac{1}{2} \times L\right) \times L \times b = \frac{1}{2} \times \left(9.8 \times 10^3 \times \frac{1}{2} \times 3\right) \times 3 \times 2 = 4.41 \times 10^4 \text{N}$$

(2) The direction of total pressure force

The resultant force F on the immersed surface is perpendicular to the plane and makes an angle of 60° with the horizontal.

(3) Acting point of total pressure force

$$L_D = \frac{2}{3}L = \frac{2}{3} \times 3 = 2\text{m}$$

Method 2 Analytic method is shown in Figure 3-22(c).

(1) The magnitude of total pressure force

$$F = p_C A = \gamma H_C L b = 9.8 \times 10^3 \times \left(\frac{3}{2} \times \frac{1}{2}\right) \times 3 \times 2 = 4.41 \times 10^4 \text{N}$$

(2) The direction of total pressure force

The resultant force F on the immersed surface is perpendicular to the plane and makes an angle of 60° with the horizontal.

(3) Acting point of total pressure force

$$y_D = \frac{I_C}{y_C A} + y_C = \left(\frac{1}{12}bL^3\right) \bigg/ \left(\frac{3}{2} \times Lb\right) + \frac{3}{2} = \frac{1}{18}L^2 + \frac{3}{2} = 2\text{m}$$

【例题 3-3】 一矩形闸门长度 $L=3$m、宽度 $b=2$m、倾角 $\alpha=30°$，如图 3-22(a) 所示，请分别用图解法与解析法求解作用在闸门上的静水总压力。

解：

方法 1：图解法，如图 3-22(b) 所示。

（1）总压力的大小

$$F = \frac{1}{2} \times \left(\gamma \times \frac{1}{2} \times L\right) \times L \times b = \frac{1}{2} \times \left(9.8 \times 10^3 \times \frac{1}{2} \times 3\right) \times 3 \times 2 = 4.41 \times 10^4 \text{N}$$

（2）总压力的方向

垂直并指向作用面，与水平方向成 60°角。

（3）总压力的作用点

$$L_D = \frac{2}{3}L = \frac{2}{3} \times 3 = 2\text{m}$$

方法 2：解析法，如图 3-22(c) 所示。

（1）总压力的大小

$$F = p_C A = \gamma H_C L b = 9.8 \times 10^3 \times \left(\frac{3}{2} \times \frac{1}{2}\right) \times 3 \times 2 = 4.41 \times 10^4 \text{N}$$

（2）总压力的方向

垂直并指向作用面，与水平方向成 60°角。

（3）总压力的作用点

$$y_D = \frac{I_C}{y_C A} + y_C = \left(\frac{1}{12}bL^3\right) \bigg/ \left(\frac{3}{2} \times Lb\right) + \frac{3}{2} = \frac{1}{18}L^2 + \frac{3}{2} = 2\text{m}$$

【Sample Problem 3-4】 As shown in Figure 3-23, both sides of a rectangular gate undergo the press of water, where $H_1 = 4.5$m, $H_2 = 2.5$m and the angle $\alpha = 45°$. Assuming the width of gate

$b=1\text{m}$, try to calculate the total pressure force acting on the gate and its acting point.

Solution: The total pressure of the liquid is the difference between total pressure force on left and that on right acting on the gate. Namely

$$F = F_1 - F_2$$

For $H_{C_1} = \dfrac{H_1}{2}$, $A_1 = bL_1 = b\dfrac{H_1}{\sin\alpha}$

$H_{C_2} = \dfrac{H_2}{2}$, $A_2 = bL_2 = b\dfrac{H_2}{\sin\alpha}$

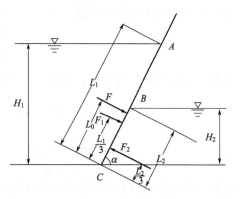

Figure 3-23　Figure for Sample Problem 3-4

图 3-23　例题 3-4 图

Thus

$$F = \rho g H_{C_1} A_1 - \rho g H_{C_2} A_2 = \frac{\rho g b H_1^2}{2\sin\alpha} - \frac{\rho g b H_2^2}{2\sin\alpha}$$

$$= \frac{9800 \times 1 \times 4.5^2}{2 \times 0.707} - \frac{9800 \times 1 \times 2.5^2}{2 \times 0.707} = 97030\text{N}$$

The coordinate of pressure center on rectangular plane

$$y_D = y_C + \frac{I_C}{y_C A} = \frac{L}{2} + \frac{bL^3/12}{(L/2)bL} = \frac{2}{3}L$$

According to the resultant moment theorem, do moment to the axis that crosses point C and keeps vertical with figure.

$$FL_0 = F_1 \frac{L_1}{3} - F_2 \frac{L_2}{3} = F_1 \frac{H_1}{3\sin\alpha} - F_2 \frac{H_2}{3\sin\alpha}$$

So

$$L_0 = \frac{F_1 H_1 - F_2 H_2}{3F\sin\alpha} = \frac{140346 \times 4.5 - 43316 \times 2.5}{3 \times 97030 \times 0.707} = 2.54\text{m}$$

This is the distance from acting point of total pressure acting on the gate to the lower end of gate.

【例题 3-4】 如图 3-23 所示，一矩形闸门两面受到水的压力，左边水深 $H_1 = 4.5\text{m}$，右边水深 $H_2 = 2.5\text{m}$，闸门与水面成 $\alpha = 45°$ 倾斜角。假设闸门的宽度 $b = 1\text{m}$，试求作用在闸门上的总压力及其作用点。

解： 作用在闸门上的总压力为左右两边液体总压力之差，即

$$F = F_1 - F_2$$

由

$$H_{C_1} = \frac{H_1}{2}, \quad A_1 = bL_1 = b\frac{H_1}{\sin\alpha}$$

$$H_{C_2} = \frac{H_2}{2}, \quad A_2 = bL_2 = b\frac{H_2}{\sin\alpha}$$

得

$$F = \rho g H_{C_1} A_1 - \rho g H_{C_2} A_2 = \frac{\rho g b H_1^2}{2\sin\alpha} - \frac{\rho g b H_2^2}{2\sin\alpha}$$

$$= \frac{9800 \times 1 \times 4.5^2}{2 \times 0.707} - \frac{9800 \times 1 \times 2.5^2}{2 \times 0.707} = 97030\text{N}$$

由于矩形平面压力中心坐标

$$y_D = y_C + \frac{I_C}{y_C A} = \frac{L}{2} + \frac{bL^3/12}{(L/2)bL} = \frac{2}{3}L$$

根据合力矩定理，对通过 C 点垂直于图面的轴取矩，得

$$FL_0 = F_1 \frac{L_1}{3} - F_2 \frac{L_2}{3} = F_1 \frac{H_1}{3\sin\alpha} - F_2 \frac{H_2}{3\sin\alpha}$$

所以

$$L_0 = \frac{F_1 H_1 - F_2 H_2}{3F\sin\alpha} = \frac{140346 \times 4.5 - 43316 \times 2.5}{3 \times 97030 \times 0.707} = 2.54 \text{m}$$

这就是作用在闸门上的总压力的作用点距闸门下端的距离。

3.7 Hydrostatic Forces on Curved Surfaces

The force acting on the curved surface in a static fluid is a complicated system of spacial forces. The questions to calculate its total force become the compound questions of system of spacial forces. Now take dualistic curved surface as an example to explain the calculation method of its total force.

As shown in Figure 3-24, assume the area of a dualistic curved surface MN is A, and its generatrix is vertical to the paper surface. The left undergoes the force of the static liquid. Take an infinitesimal area dA on the curved surface. The submerged depth at the centroid is h. Then the total force of fluid acting on the infinitesimal area is

$$dF = \rho g h \, dA$$

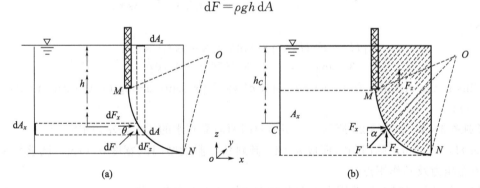

Figure 3-24 Hydrostatic Force on the Curved Surface
图 3-24 作用在曲面上的静水总压力

3.7 作用在曲面上的静水总压力

静止流体作用在曲面上的压力为一复杂的空间力系，求其总压力的问题便成为空间力系的合成问题。下面以工程中常见的二元曲面为例，说明确定其总压力的计算方法。

设有一面积为 A 的二元曲面 MN，其母线垂直于纸面，左侧承受静止液体压力的作用，如图 3-24 所示，在曲面上任取一微元面积 dA，其形心点的淹没深度为 h，则流体作用在微元上的总压力为

$$dF = \rho g h \, dA$$

3.7.1 The Magnitude of Total Pressure Force

(1) The horizontal force Assume the normal of infinitesimal area dA and axis x forms an angle θ, and then the horizontal component of the infinitesimal area is
$$dF_x = \rho g h \, dA \cos\theta = \rho g h \, dA_x$$
So the horizontal component of the total pressure force is
$$F_x = \rho g \int_A h \, dA_x = \rho g h_C A_x \tag{3-42}$$
where A_x —the area of the projection on plane yoz of area A of curved surface;

h_C —the submerged depth of the centroid of A_x.

The horizontal force acting on the curved surface equals to the total force of fluid acting on the vertical projection plane A_x of the curved surface.

(2) The vertical force The vertical force acting on the infinitesimal area is
$$dF_z = \rho g h \, dA \sin\theta = \rho g h \, dA_z$$
So the vertical force is
$$F_z = \rho g \int_A h \, dA_z \tag{3-43}$$
where $\int_A h \, dA_z$ —the volume of liquid column on curved surface MN, is called pressure volume, marked as V.

So
$$F_z = \rho g V \tag{3-44}$$
The vertical component F_z of the total pressure force acting on the curved surface MN equals to the liquid weight of its pressure volume.

(3) Total pressure force Total pressure force is
$$F = \sqrt{F_x^2 + F_z^2} \tag{3-45}$$

3.7.1 总压力的大小

(1) 总压力的水平分力 设 θ 为微元面积 dA 的法线与 x 轴的夹角，则微元水平分力
$$dF_x = \rho g h \, dA \cos\theta = \rho g h \, dA_x$$
故总压力的水平分力为
$$F_x = \rho g \int_A h \, dA_x = \rho g h_C A_x \tag{3-42}$$
式中，A_x 为曲面面积 A 在 yoz 平面上的投影面积；h_C 为 A_x 的形心点的淹没深度。

流体作用在曲面上总压力的水平分力等于流体作用在该曲面的铅垂投影面 A_x 上的总压力。

(2) 总压力的垂直分力 作用在微元上的垂直分力为
$$dF_z = \rho g h \, dA \sin\theta = \rho g h \, dA_z$$
故总压力的垂直分力为
$$F_z = \rho g \int_A h \, dA_z \tag{3-43}$$
式中，$\int_A h \, dA_z$ 为曲面 MN 上的液柱体积，称为压力体，记作 V。

故
$$F_z = \rho g V \tag{3-44}$$

作用在曲面 MN 上的总压力的垂直分力 F_z 等于其压力体的液重。

（3）总压力 总压力为
$$F=\sqrt{F_x^2+F_z^2} \tag{3-45}$$

3.7.2 The Direction of Total Pressure Force

The direction of total pressure force can be represented with the angle α which it makes with the horizontal

$$\alpha=\arctan\frac{F_z}{F_x} \tag{3-46}$$

3.7.2 总压力的方向

总压力的作用方向用其与水平方向的交角 α 来表示。

$$\alpha=\arctan\frac{F_z}{F_x} \tag{3-46}$$

3.7.3 The Acting Point of Total Pressure Force

The acting line of vertical force passes the center of gravity of pressure volume, the acting line of horizontal force passes the pressure center of A_x, and both of them point to the surface. Therefore, the acting line of total force necessarily passes the crossing point of acting lines and makes an angle of α with the vertical (Figure 3-24). The crossing point is the acting point on the curved surface of total force.

3.7.3 总压力的作用点

由于总压力的垂直分力作用线通过压力体的重心，水平分力的作用线通过 A_x 的压力中心，且二者均指向受压面，故总压力作用线必通过这两条作用线的交点，且与垂线成 α 角（见图 3-24）。这条总压力的作用线与曲面的交点就是总压力在曲面上的作用点。

3.7.4 Pressure Volume

The pressure volume is the geometric volume determined by the integral equation $\int_A h\,dA_z$. It has no relation with that whether there is a liquid in it or not. The liquid weight of pressure volume is not always the factual gravity of liquid in the pressure volume. There are usually three forms: true pressure volume, empty pressure volume and compound pressure volume (as shown in Figure 3-25).

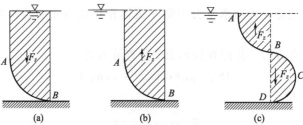

Figure 3-25 Pressure Volume on Curved Surface
(a) True pressure volume; (b) Empty pressure volume; (c) Compound pressure volume

图 3-25 曲面上的压力体
(a) 实压力体；(b) 虚压力体；(c) 混合压力体

3.7.4 压力体

压力体是由积分式 $\int_A h\,\mathrm{d}A_z$ 所确定的纯几何体积，它与这块体积中究竟有无液体没有关系。压力体的液重并不一定是压力体内实际具有的液体重力，压力体的形式通常有实压力体、虚压力体及混合压力体三种（图 3-25）。

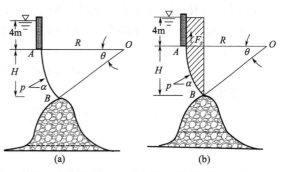

Figure 3-26　Figure for Sample Problem 3-5

图 3-26　例题 3-5 图

【Sample Problem 3-5】 As shown in Figure 3-26（a）, there is a camber gate whose width b is 8m and radius R is 10m. The left side is filled with water. Given that $\theta=30°$, calculate the total pressure force acting on the gate and draw a picture of pressure volume.

Solution：

(1) The magnitude of total pressure force

① The horizontal force

$$F_x = \gamma h_C A_x = \gamma(4+H/2)Hb$$
$$= 9800 \times (4+10\times\sin30°/2)\times 10\times\sin30°\times 8 = 2548.0\,\mathrm{kN}$$

② The vertical force

As shown in Figure 3-26(b), a picture of pressure volume is drawn and we will calculate its volume.

$$V = [4\times(R-R\cos30°)+(\pi R^2\times 30/360 - R\sin30°\times R\cos30°/2)]\times b$$
$$= [4\times(10-10\times\cos30°)+(\pi\times 10^2\times 30/360 - 10\times\sin30°\times 10\times\cos30°/2)]\times 8$$
$$= 79.1\,\mathrm{m}^3$$

$$F_z = \gamma V = 9800\times 79.1 = 775.2\,\mathrm{kN}$$

③ The magnitude of total pressure force

$$F = \sqrt{F_x^2 + F_z^2} = \sqrt{2548.0^2 + 775.2^2} = 2663.3\,\mathrm{kN}$$

(2) The direction of total pressure force

$$\alpha = \arctan\frac{F_z}{F_x} = \arctan\frac{775.2}{2548.0} = 16.9°$$

(3) The acting point of total pressure force

$$4+R\sin16.9° = 4+10\sin16.9° = 6.91\,\mathrm{m}$$

The distance between water surface and acting point is 6.91m.

【例题 3-5】 图 3-26(a) 所示为一溢流坝上的弧形闸门。已知，$R=10\mathrm{m}$，门宽 $b=8\mathrm{m}$，$\theta=30°$，左侧充满水。试求作用在该弧形闸门上的静水总压力，并在图中画出压力体。

解：

(1) 总压力的大小

① 水平方向分力

$$F_x = \gamma h_C A_x = \gamma(4+H/2)Hb = 9800\times(4+10\times\sin30°/2)\times 10\times\sin30°\times 8 = 2548.0\,\mathrm{kN}$$

② 铅垂方向分力

在图 3-26(b) 中绘出压力体图，压力体体积 V 为

$$V = [4 \times (R - R\cos 30°) + (\pi R^2 \times 30/360 - R\sin 30° \times R\cos 30°/2)] \times b$$
$$= [4 \times (10 - 10 \times \cos 30°) + (\pi \times 10^2 \times 30/360 - 10 \times \sin 30° \times 10 \times \cos 30°/2)] \times 8$$
$$= 79.1 \text{ m}^3$$
$$F_z = \gamma V = 9800 \times 79.1 = 775.2 \text{ kN}$$

③ 总压力
$$F = \sqrt{F_x^2 + F_z^2} = \sqrt{2548.0^2 + 775.2^2} = 2663.3 \text{ kN}$$

(2) 方向
$$\alpha = \arctan \frac{F_z}{F_x} = \arctan \frac{775.2}{2548.0} = 16.9°$$

(3) 作用点
$$4 + R\sin 16.9° = 4 + 10\sin 16.9° = 6.91 \text{ m}$$

因此作用点位于水面下 6.91m 处。

Exercises 习题

3-1 The open ends of a glass tube in the closed container filled with water are shown in Figure. Given that when the glass tube extends the depth $h = 1.5$m under water surface, no air flows into the container through the glass tube and no water flows into the glass tube. Try to calculate the absolute pressure and gage pressure on the water surface in the container at this time.

3-1 封闭盛水容器中的玻璃管两端开口，如图所示，已知玻璃管伸入水面以下 $h = 1.5$m 时，既无空气通过玻璃管进入容器，又无水进入玻璃管，试求此时容器内水面上的绝对压强和相对压强。

3-2 In Figure, pressure gage A reads 1.5kPa (gage). The fluids are at 20℃. Determine the elevations z, in meters, of the liquid levels in the open piezometer tubes B and C.

3-2 在图中，压力表 A 的读数为 1.5kPa，流体温度为 20℃，求在敞开的测压管 B 和 C 中液面的高度，以米计。

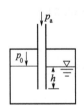

Figure for Exercise 3-1

习题 3-1 图

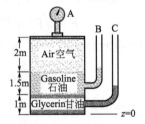

Figure for Exercise 3-2

习题 3-2 图

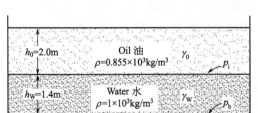

Figure for Exercise 3-3

习题 3-3 图

3-3 An open tank contains water 1.40m deep covered by a 2-m-thick layer of oil ($s = 0.855$). What is the pressure head at the bottom of the tank, in terms of a water column?

3-3 如图所示，一个敞口的水槽蓄水 1.40m 深，且表面覆盖了 2m 厚的油层（$\rho = 0.855 \times 10^3$kg/m³），用水柱表示，水槽底部的压头是多少？

3-4 In Figure both the tank and the tube are open to the atmosphere. The specific gravities (SG) of oil and water are 0.8 and 1.0, respectively. If $L=2.5$m, what is the angle of tilt θ of the tube?

3-4 在图中，容器和管子都是敞开的，油和水的相对密度分别为0.8和1.0。如果$L=2.5$m，管子的倾斜角θ是多少？

3-5 The upside of a micro-manometer of cup type is filled with oil, $\gamma_{oil}=9.0$kN/m^3, and the downside is filled with water. The diameter of the cup is $D=40$mm and the diameter of the tube is $d=4$mm. If $p_2-p_1=10$mm H$_2$O, what is the water difference h?

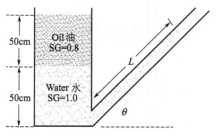

Figure for Exercise 3-4
习题 3-4 图

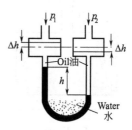

Figure for Exercise 3-5
习题 3-5 图

3-5 杯式微压计如图所示，上部盛油，$\gamma_{oil}=9.0$kN/m^3，下部为水，圆杯直径$D=40$mm，圆管直径$d=4$mm，若接入压力$p_2-p_1=10$mm的水柱，水位差h应为多少？

3-6 The tank of water in Figure accelerates uniformly by freely rolling down a 30° incline. If the wheels are frictionless, what is the angle θ? Can you explain this interesting result?

3-6 如图所示的水箱从倾角为30°的斜面上自由滚落下来。如果轮子是光滑的，则角度θ是多少？你能解释这个有趣的结果吗？

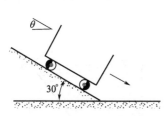

Figure for Exercise 3-6
习题 3-6 图

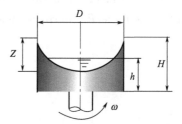

Figure for Exercise 3-7
习题 3-7 图

3-7 A cylindrical rotating container is filled with some water. Given that $D=300$mm, $H=500$mm and $h=300$mm, what is the rotational speed ω when the water surface just reaches the upper margin?

3-7 如图所示，一圆柱形旋转容器，装部分水，已知$D=300$mm、$H=500$mm、$h=300$mm，转速ω为多少时，水面恰好达到容器的上缘？

3-8 A cylindrical container is filled with water. An open manometric tube is installed at the coping center. What is the upward total pressure acting on the coping when the container rotates at the angular speed ω?

3-8　如图所示，一圆柱形容器装满水，顶盖中心装有敞口测压管，当以角速度 ω 旋转时，顶盖受到多大的向上液体的总压力？

3-9　A cheap accelerometer, probably worth the price, can be made from a U-tube as shown in Figure. If $L=18$cm and $D=5$mm, what will h be if $a_x=6$m/s^2? Can the scale markings on the tube be linear multiples of a_x?

3-9　如图所示，加速度可以用 U 形管测得。已知 $L=18$cm、$D=5$mm，如果 $a_x=6$m/s^2，这时 h 是多少？这个 U 形管上的标注能表示加速度 a_x 的大小吗？

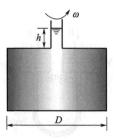

Figure for Exercise 3-8
习题 3-8 图

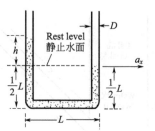

Figure for Exercise 3-9
习题 3-9 图

3-10　In Figure, the hydrostatic force F is the same on the bottom of all three containers, even though the weights of liquid above are quite different. The three bottom shapes and the liquids are the same. This is called the hydrostatic paradox. Please explain why it is true.

3-10　在图中，三个容器底部的静水压力相同，而里面盛的水的重量完全不同。这三个容器底部的形状和内部装的液体都是相同的，这就是所谓的静水奇象，请解释一下这一事实。

3-11　Gate AB has length L and width b into the paper, is hinged at B, and has negligible weight. The liquid level h remains at the top of the gate for any angle θ. Find an analytic expression for the force P, perpendicular to AB, required to keep the gate in equilibrium in Figure.

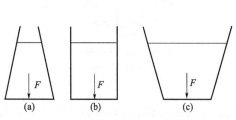

Figure for Exercise 3-10
习题 3-10 图

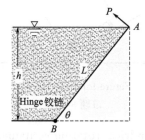

Figure for Exercise 3-11
习题 3-11 图

3-11　如图所示，闸门 AB 长度为 L，宽度为 b，铰接在 B 点，自重忽略不计，不管闸门处于什么角度，液面高度 h 始终在闸门的最高点。为了保持闸门的平衡，试求垂直作用于 AB 的力 P 的解析表达式。

3-12　Panel BC in Figure is circular. Calculate: (a) the hydrostatic force of the water on the panel, (b) its center of pressure, and (c) the moment of this force about point B.

3-12 图中板 BC 是圆形的。计算：(a) 该板上的静水压力；(b) 压强的中心点；(c) 该力关于 B 点的力矩。

3-13 (a) Find the horizontal and vertical forces per meter of width acting on the camber gate in Figure; (b) locate the horizontal force and indicate the line of action of the vertical force without actually computing its location.

3-13 如图所示，(a) 求作用在弧形闸门上水平和铅垂方向上每米宽的力；(b) 找出水平力和铅垂方向的作用线，不需要求出其实际位置。

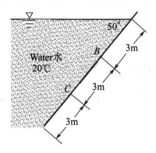

Figure for Exercise 3-12

习题 3-12 图

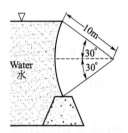

Figure for Exercise 3-13

习题 3-13 图

3-14 As shown in Figure, there is a cylindrical rotating gate whose length is 1m and diameter D is 4m. Both sides are filled with water. Given that the depths of water on the upstream and downstream sides are 4m and 2m, respectively. Calculate the total force acting on the gate and its direction.

3-14 如图所示，有一圆形滚门，长 1m，直径 D 为 4m，两侧有水，上游水深 4m，下游水深 2m，求作用在门上的总压力的大小及作用线的位置。

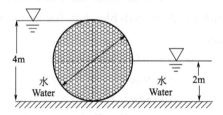

Figure for Exercise 3-14

习题 3-14 图

Chapter 4 Fluid Flow and Basic Equations
第 4 章 流体运动及其基本方程

4.1 Lagrange's Method and Euler's Method

The main methods to describe the fluid motion are Lagrange's method and Euler's method.

4.1 拉格朗日法和欧拉法

描述流体运动的方法主要有拉格朗日法和欧拉法。

4.1.1 Lagrange's Method

Lagrange's method is to consider the fluid particles as research objects and to research the motion course of each particle, and then obtain the kinetic regulation of the total flow through synthesizing motion instances of all the objects that are researched. Lagrange's method is a method of particle coordinates.

When we use Lagrange's method to describe the fluid motion, the position coordinates of moving particles are not independent variables but functions of original coordinate a, b, c and time variable t, namely

$$\left. \begin{array}{l} x = x(a, b, c, t) \\ y = y(a, b, c, t) \\ z = z(a, b, c, t) \end{array} \right\} \quad (4\text{-}1)$$

In this equation, a, b, c and t are all called Lagrangian variables. Different particles have different original coordinates. Some difficulties in mathematics will be met when using Lagrange's method to analyze fluid motion except for few instances, such as research on wave motion. Euler's method is used mostly to study the fluid motion.

4.1.1 拉格朗日法

把流体质点作为研究对象，研究各质点的运动历程，然后通过综合所有被研究流体质点的运动情况来获得整个流体运动的规律，这种方法叫作拉格朗日法，实质上是一种质点系法。

用拉格朗日法描述流体的运动时，运动质点的位置坐标不是独立变量，而是起始坐标 a、b、c 和时间变量 t 的函数，即

$$\left. \begin{array}{l} x = x(a, b, c, t) \\ y = y(a, b, c, t) \\ z = z(a, b, c, t) \end{array} \right\} \quad (4\text{-}1)$$

式中，a、b、c、t 统称拉格朗日变量，不同的运动质点，起始坐标不同。

用拉格朗日法分析流体运动，除少数情况外，如研究波浪运动，在数学上将会遇到困难。在研究流体运动时多采用欧拉法。

4.1.2 Euler's Method

Without researching the moving course of each particle, but with a view to the space points in the fluid field (the space full of moving fluid), the regulations of motion factors with respect to time are observed when the particle flows over each space point, and then the regulation of the total flow is gained by synthesizing the motion regulations of enough space points. This is called Euler's method (fluid field method).

When we use Euler's method to describe fluid motion, the motion factors are continuous differential functions of space coordinates x, y, z and time variable t. x, y, z and t are called Euler's variables. So the velocity field can be expressed by the following equation:

$$\left.\begin{array}{l} u_x = u_x(x,y,z,t) \\ u_y = u_y(x,y,z,t) \\ u_z = u_z(x,y,z,t) \end{array}\right\} \tag{4-2}$$

Pressure field and density field can be expressed as:

$$p = p(x,y,z,t) \tag{4-3}$$

$$\rho = \rho(x,y,z,t) \tag{4-4}$$

In Equation (4-2), x, y and z are motion coordinates of fluid particles at time t, namely they are functions of time t. According to the principle of compound function differentiation and the following equations:

$$\frac{\mathrm{d}x}{\mathrm{d}t} = u_x, \quad \frac{\mathrm{d}y}{\mathrm{d}t} = u_y, \quad \frac{\mathrm{d}z}{\mathrm{d}t} = u_z$$

the acceleration components in the x, y and z directions are:

$$\left.\begin{array}{l} a_x = \dfrac{\mathrm{d}u_x}{\mathrm{d}t} = \dfrac{\partial u_x}{\partial t} + u_x \dfrac{\partial u_x}{\partial x} + u_y \dfrac{\partial u_x}{\partial y} + u_z \dfrac{\partial u_x}{\partial z} \\ a_y = \dfrac{\mathrm{d}u_y}{\mathrm{d}t} = \dfrac{\partial u_y}{\partial t} + u_x \dfrac{\partial u_y}{\partial x} + u_y \dfrac{\partial u_y}{\partial y} + u_z \dfrac{\partial u_y}{\partial z} \\ a_z = \dfrac{\mathrm{d}u_z}{\mathrm{d}t} = \dfrac{\partial u_z}{\partial t} + u_x \dfrac{\partial u_z}{\partial x} + u_y \dfrac{\partial u_z}{\partial y} + u_z \dfrac{\partial u_z}{\partial z} \end{array}\right\} \tag{4-5}$$

The vector of acceleration can be expressed as

$$\vec{a} = \frac{\mathrm{d}\vec{u}}{\mathrm{d}t} = \frac{\partial \vec{u}}{\partial t} + (\vec{u} \cdot \nabla)\vec{u}$$

In this equation

$$\nabla = \vec{i}\frac{\partial}{\partial x} + \vec{j}\frac{\partial}{\partial y} + \vec{k}\frac{\partial}{\partial z}$$

Acceleration consists of temporal acceleration and spatial acceleration. Temporal acceleration $\dfrac{\partial \vec{u}}{\partial t}$ denotes the variety of velocity of fluid particles through fixed space points with respect to time. Spatial acceleration $(\vec{u} \cdot \nabla)\vec{u}$ denotes variational ratio of velocity with respect to space.

The general formula for solving variational ratio of other motion factors with respect to time is

$$\frac{d}{dt} = \frac{\partial}{\partial t} + (\vec{u} \cdot \nabla) \tag{4-6}$$

where $\dfrac{d}{dt}$ —— total derivative;

$\dfrac{\partial}{\partial t}$ —— temporal derivative;

$(\vec{u} \cdot \nabla)$ —— spatial derivative.

4.1.2 欧拉法

不研究各个质点的运动过程，而着眼于流场（充满运动流体的空间）中的空间点，通过观察质点流经每个空间点时的运动要素随时间变化的规律，把足够多的空间点综合起来而得出整个流体运动的规律，这种方法叫作欧拉法（流场法）。

用欧拉法描述流体的运动时，运动要素是空间坐标 x、y、z 和时间变量 t 的连续可微函数。x、y、z、t 称为欧拉变量，因此速度场可表示为：

$$\left.\begin{array}{l} u_x = u_x(x,y,z,t) \\ u_y = u_y(x,y,z,t) \\ u_z = u_z(x,y,z,t) \end{array}\right\} \tag{4-2}$$

压强和密度场表示为：

$$p = p(x,y,z,t) \tag{4-3}$$

$$\rho = \rho(x,y,z,t) \tag{4-4}$$

式中，x、y、z 是流体质点在 t 时刻的运动坐标，即它们是时间变量 t 的函数。因此，根据复合函数求导法则，并考虑到

$$\frac{dx}{dt} = u_x, \quad \frac{dy}{dt} = u_y, \quad \frac{dz}{dt} = u_z$$

可得加速度在空间坐标 x、y、z 方向的分量为

$$\left.\begin{array}{l} a_x = \dfrac{du_x}{dt} = \dfrac{\partial u_x}{\partial t} + u_x \dfrac{\partial u_x}{\partial x} + u_y \dfrac{\partial u_x}{\partial y} + u_z \dfrac{\partial u_x}{\partial z} \\[6pt] a_y = \dfrac{du_y}{dt} = \dfrac{\partial u_y}{\partial t} + u_x \dfrac{\partial u_y}{\partial x} + u_y \dfrac{\partial u_y}{\partial y} + u_z \dfrac{\partial u_y}{\partial z} \\[6pt] a_z = \dfrac{du_z}{dt} = \dfrac{\partial u_z}{\partial t} + u_x \dfrac{\partial u_z}{\partial x} + u_y \dfrac{\partial u_z}{\partial y} + u_z \dfrac{\partial u_z}{\partial z} \end{array}\right\} \tag{4-5}$$

矢量式为

$$\vec{a} = \frac{d\vec{u}}{dt} = \frac{\partial \vec{u}}{\partial t} + (\vec{u} \cdot \nabla)\vec{u}$$

其中

$$\nabla = \vec{i}\frac{\partial}{\partial x} + \vec{j}\frac{\partial}{\partial y} + \vec{k}\frac{\partial}{\partial z}$$

加速度包括时变加速度和位变加速度。时变加速度 $\dfrac{\partial \vec{u}}{\partial t}$ 表示通过固定空间点的流体质点速度随时间的变化。位变加速度 $(\vec{u} \cdot \nabla)\vec{u}$ 表示流体质点所在空间位置的变化所引起的速度变化率。

用欧拉法求流体质点其他运动要素对时间变化率的一般式子为

$$\frac{d}{dt}=\frac{\partial}{\partial t}+(\vec{u}\cdot\nabla) \tag{4-6}$$

式中，$\frac{d}{dt}$ 为全导数；$\frac{\partial}{\partial t}$ 为时变导数；$(\vec{u}\cdot\nabla)$ 为位变导数。

4.1.3 Path Line and Streamline

(1) Path line When Lagrange's method is used to describe fluid motion, the concept of path line should be introduced. A path line is the trajectory of a fluid particle. The trajectory of a certain fluid particle moving with time is shown in Figure 4-1.

For $ds = u dt$, the differential equation of path line is

$$\frac{dx}{u_x(x,y,z,t)}=\frac{dy}{u_y(x,y,z,t)}=\frac{dz}{u_z(x,y,z,t)}=dt \tag{4-7}$$

(2) Streamline When Euler's method is used to describe fluid motion vividly, the concept of streamline should be introduced.

A streamline is a continuous line drawn through the fluid so that it has the direction of the velocity vector at any point. As shown in Figure 4-2, streamline is a line whose tangent always represents the direction of velocity.

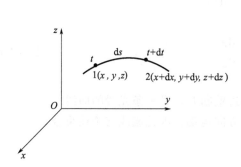

Figure 4-1　Path Line
图 4-1　迹线

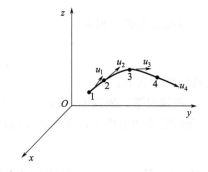

Figure 4-2　Streamline
图 4-2　流线

Suppose the velocity vector of a certain point on streamline is

$$\vec{u}=u_x\vec{i}+u_y\vec{j}+u_z\vec{k}$$

The element segment vector $\vec{ds}$ on streamline is

$$\vec{ds}=dx\vec{i}+dy\vec{j}+dz\vec{k}$$

According to the definition of streamline, the differential equation expressed by vector is

$$\vec{u}\times\vec{ds}=\begin{vmatrix} \vec{i} & \vec{j} & \vec{k} \\ u_x & u_y & u_z \\ dx & dy & dz \end{vmatrix}=0 \tag{4-8(a)}$$

If Equation [4-8(a)] is expressed by projection form, then the equation changes into

$$\frac{dx}{u_x}=\frac{dy}{u_y}=\frac{dz}{u_z} \tag{4-8(b)}$$

The characteristics of streamlines are described as follows:

① In general, streamlines will not intercross and will not end at a solid wall. They must be smoothed curves.

② In steady flow, a particle always moves tangent to the streamline. Hence the path line of a particle is a streamline.

4.1.3 迹线和流线

（1）迹线　用拉格朗日法描述流体运动，需要引进迹线的概念。迹线是流体质点运动的轨迹线，图 4-1 为某流体质点随时间推移所走的轨迹。根据 $\mathrm{d}s = u\mathrm{d}t$，迹线微分方程为

$$\frac{\mathrm{d}x}{u_x(x,y,z,t)} = \frac{\mathrm{d}y}{u_y(x,y,z,t)} = \frac{\mathrm{d}z}{u_z(x,y,z,t)} = \mathrm{d}t \tag{4-7}$$

（2）流线　用欧拉法形象地对流场进行几何描述，引进了流线的概念。某一瞬时在流场中绘出的曲线，在这条曲线上所有质点的速度矢量都和该曲线相切，则称此曲线为流线。流线上某点的切线方向代表该点的速度方向，如图 4-2 所示。设流线上一点的速度矢量为 $\vec{u} = u_x\vec{i} + u_y\vec{j} + u_z\vec{k}$，流线上的微元线段矢量 $\mathrm{d}\vec{s} = \mathrm{d}x\vec{i} + \mathrm{d}y\vec{j} + \mathrm{d}z\vec{k}$，根据流线定义，可得用矢量表示的微分方程为

$$\vec{u} \times \mathrm{d}\vec{s} = \begin{vmatrix} \vec{i} & \vec{j} & \vec{k} \\ u_x & u_y & u_z \\ \mathrm{d}x & \mathrm{d}y & \mathrm{d}z \end{vmatrix} = 0 \tag{4-8(a)}$$

若写成投影形式，则为

$$\frac{\mathrm{d}x}{u_x} = \frac{\mathrm{d}y}{u_y} = \frac{\mathrm{d}z}{u_z} \tag{4-8(b)}$$

流线的性质如下：
① 一般情况下，流线不能相交，不会在固体边壁断开，是一条光滑的曲线。
② 在恒定流动条件下，质点沿着流线的切线方向运动，因此流线与迹线重合。

4.2 Basic Concepts of Fluid Flow

4.2.1 Stream Tube, Stream Flow, Element Flow, Total Flow and Cross Section of Flow

(1) Stream tube　As shown in Figure 4-3, a stream tube is the tube made by all streamlines passing through a small closed curve. In steady flow it is fixed in space and can have no flow through its walls because the velocity vector has no component normal to the tube surface.

(2) Stream flow　As shown in Figure 4-3, the summation of all streamlines in stream tube is called stream flow.

(3) Element flow　As shown in Figure 4-4, the stream whose sections are infinitesimal is called element flow.

(4) Total flow　As shown in Figure 4-4, the summation of countless element flows is called total flow.

(5) Cross section of flow　As shown in Figure 4-5, the section which keeps orthogonal with all the streamlines in the stream tube are called cross section of flow. When all streamlines keep parallel, the cross section is a plane. Or else the cross section is a curve surface.

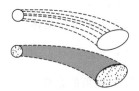
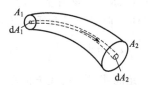
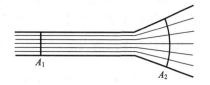

Figure 4-3 Stream Tube and Stream Flow

图 4-3 流管和流束

Figure 4-4 Element Flow and Total Flow

图 4-4 元流和总流

Figure 4-5 Cross Section of Flow

图 4-5 过流断面

4.2 流体运动的基本概念

4.2.1 流管、流束、元流、总流、过流断面

（1）流管 在流场中取任意封闭的曲线，经过曲线的每一点做流线，由这些流线所围成的管称为流管，如图 4-3 所示。在恒定流中，速度没有垂直于流管表面的分量，因此流管的面是固定的，且没有流体穿过它的面。

（2）流束 如图 4-3 所示，流管内所有流线的总和称为流束。

（3）元流 如图 4-4 所示，过流断面面积无限小的流束称为元流。

（4）总流 如图 4-4 所示，总流可以看成是由流动边界内无数元流所组成的总和。

（5）过流断面 如图 4-5 所示，与流束中所有流线正交的横断面称为过流断面。当组成流束的所有流线互相平行时，过流断面是平面；否则，过流断面是曲面。

4.2.2 Flow Rate and Mean Velocity

(1) Flow rate The flow rate is the quantity of fluid flowing per unit time across any section. There are three kinds of expressing methods as following.

① The method which is expressed by the fluid volume in unit time is called volume flow rate or discharge. That is Q (m^3/s or L/s).

② The method which is expressed by the fluid mass in unit time is called mass flow rate. That is Q_m (kg/s).

③ The method which is expressed by the fluid weight in unit time is called weight flow rate. That is Q_w (kN/s).

Commonly, volume flow rate is expressed for incompressible fluids. Mass flow rate and weight flow rate are expressed for compressible fluids.

The discharge flowing via the random curved surface is

$$Q = \int_A \vec{u} \cdot \vec{n} \, dA = \int_A u \cos(\vec{u}, \vec{n}) dA$$

In this equation, $\cos(\vec{u}, \vec{n})$ is the cosine of inclination of velocity vector and the unit vector $\vec{n}$ in normal orientation of infinitesimal area dA.

4.2.2 流量与断面平均速度

（1）流量 流量是指在单位时间内通过过流断面的流体数量，有下列三种表示方法。

① 以单位时间通过的流体体积表示，称为体积流量（流量），记作 Q（m^3/s 或 L/s）。

② 以单位时间通过的流体质量表示，称为质量流量，记作 Q_m（kg/s）。

③ 以单位时间通过的流体重量表示，称为重量流量，记作 Q_w（kN/s）。

通常体积流量用来表示不可压缩流体，质量流量和重量流量用来表示可压缩流体。

流经任意曲面的流量
$$Q = \int_A \vec{u} \cdot \vec{n} \, \mathrm{d}A = \int_A u \cos(\vec{u}, \vec{n}) \, \mathrm{d}A$$

式中，$\cos(\vec{u}, \vec{n})$ 为速度矢量与微元面积 $\mathrm{d}A$ 法线方向单位矢量 $\vec{n}$ 夹角的余弦。

(2) Mean velocity As shown in Figure 4-6, mean velocity is the discharge per unit area passing through the cross section. It can be calculated by the following equation:

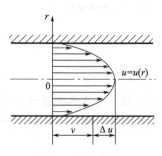

Figure 4-6 Mean Velocity of Section
图 4-6 断面平均流速

$$v = \frac{Q}{A} = \frac{\int_A u \, \mathrm{d}A}{A}$$

where v—mean velocity, m/s;
Q—discharge, m³/s;
A—the area of the cross section, m²;
u—the velocity passing through infinitesimal area $\mathrm{d}A$, m/s.

(2) 断面平均流速 如图 4-6 所示，通过某一过流断面单位面积上的体积流量称为断面平均速度，可按下式计算：

$$v = \frac{Q}{A} = \frac{\int_A u \, \mathrm{d}A}{A}$$

式中，v 为断面平均流速，m/s；Q 为体积流量，m³/s；A 为过流断面面积，m²；u 为通过微元面积 $\mathrm{d}A$ 的流速，m/s。

4.3 Types of Fluid Flow

4.3.1 Steady Flow and Unsteady Flow

Steady flow means steady with respect to time. Thus all properties of the flow at a point remain constant with respect to time. Namely

$$\frac{\partial u}{\partial t} = 0, \quad \frac{\partial p}{\partial t} = 0, \quad \frac{\partial \rho}{\partial t} = 0$$

In a unsteady flow, the flow properties at a point will change with time. Namely

$$\frac{\partial u}{\partial t} \neq 0, \quad \frac{\partial p}{\partial t} \neq 0, \quad \frac{\partial \rho}{\partial t} \neq 0$$

4.3 流体运动的类型

4.3.1 恒定流与非恒定流

恒定流意味着随时间很稳定，流场中各空间点上的一切运动要素都不随时间变化。即

$$\frac{\partial u}{\partial t} = 0, \quad \frac{\partial p}{\partial t} = 0, \quad \frac{\partial \rho}{\partial t} = 0$$

对于非恒定流，流场中各空间点上的一切运动要素都随时间变化。即

$$\frac{\partial u}{\partial t} \neq 0, \quad \frac{\partial p}{\partial t} \neq 0, \quad \frac{\partial \rho}{\partial t} \neq 0$$

4.3.2 Uniform Flow and Non-Uniform Flow

When both size and shape of cross-section are constant along the stream, we say the flow is uniform. The streamlines of a uniform flow are necessarily straight lines and parallel to each other.

Mathematical representations of uniform flow and non-uniform flow are:

Uniform flow $\qquad \dfrac{\partial p}{\partial l}=0, \quad \dfrac{\partial u}{\partial l}=0$

Non-uniform flow $\qquad \dfrac{\partial p}{\partial l}\neq 0, \quad \dfrac{\partial u}{\partial l}\neq 0$

4.3.2 均匀流与非均匀流

当过流断面的尺寸与形状沿流动方向不变时，这样的流动称为均匀流，均匀流的流线是直线并且相互平行。

均匀流与非均匀流的数学表达式为：

均匀流 $\qquad \dfrac{\partial p}{\partial l}=0, \quad \dfrac{\partial u}{\partial l}=0$

非均匀流 $\qquad \dfrac{\partial p}{\partial l}\neq 0, \quad \dfrac{\partial u}{\partial l}\neq 0$

4.3.3 Gradually-Varying Flow and Rapidly-Varying Flow

In Figure 4-7, we can classify non-uniform flows into gradually-varying flows and rapidly-varying flows. The characteristics of gradually-varying flow are that curvatures of all streamlines are small and streamlines are nearly parallel. Or else it is called rapidly-varying flow.

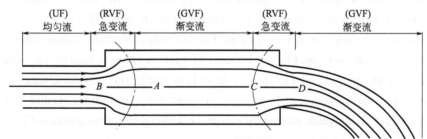

Figure 4-7　Uniform Flow and Non-uniform Flow
图 4-7　均匀流与非均匀流

UF—Uniform Flow 均匀流；GVF—Gradually-varying Flow 渐变流；RVF—Rapidly-varying Flow 急变流

4.3.3 渐变流和急变流

在图 4-7 中，非均匀流可以分为渐变流和急变流，渐变流的特点是流线的曲率很小，流线近似平行，否则就是急变流。

4.3.4 Pressure Flow, Gravity Flow and Shooting Flow

Pressure flow implies that flow occurs under pressure. Gases always flow in this manner. When a liquid flows with a free surface (for example, a partly full pipe flow), we refer to the flow as gravity flow. Liquids also flow without a free surface under pressure (for example, a full pipe flow). We refer to the fluid flowing around air or liquid space as shooting flow (for example, an orifice flow).

4.3.4 有压流、重力流及射流

有压流是在有压力的情况下发生的，气体的流动一般是这种形式。具有自由表面的液体流动称为重力流（比如非满流的管道流动）。无自由表面的液体也能进行有压流动（比如满流的管道流动）。围绕大气或液体的流体运动，我们称之为射流（比如孔口的流动）。

4.3.5 One-, Two-, and Three-dimensional Flow

The flow of which motion factor is the function of a coordinate is called one-dimensional flow; the flow of which motion factor is the function of two coordinates is called two-dimensional flow; and the flow of which motion factor is the function of three coordinates is called three-dimensional flow.

4.3.5 一维流动、二维流动及三维流动

运动要素是一个坐标的函数，这种流动称为一维流动。
运动要素是两个坐标的函数，这种流动称为二维流动。
运动要素是三个坐标的函数，这种流动称为三维流动。

4.4 System and Control Volume

The fluid mechanics system means the fluid group consisting of the determinate fluid particles. All out of the system are called outside. The real or ostensible surface that divides the system and outside is called the border of the system. The border of system moves with fluid and the volume of system, the shape and magnitude of the border surface can change with time. On the border of system there is not mass exchange, that is to say, no borders for fluid to flow in or flow out of the system. The borders of system endure the surface force that outside acts on the system. On the borders of system there is energy exchange, that is to say, there are borders for energy to flow in or flow out of the system. Using system to research fluid motion means to adopt Lagrange's viewpoints. In fluid mechanics Euler's method is used to research fluid motion commonly except for some specific circumstances. Corresponding to this, the concept of control volume is imported.

Relative to any coordinate system, the changeless volume of any space in which there are fluids to flow through is called control volume. The border surfaces of control volume are called control faces. They are always closed surfaces. The characters of control faces are as follows:

① The control faces relative to coordinate system are changeless;

② On control faces there are mass exchanges, that is to say, fluids can flow in or flow out of control volume;

③ On control faces other objects except control volume act on the fluid in control volume;

④ On control faces there are energy exchanges, that is to say, energy can import in or export through control faces.

4.4 系统与控制体

在流体力学中，系统是指由确定的流体质点所组成的流体团。系统以外的一切统称为外界。将系统和外界分开的真实或假想的表面称为系统的边界。系统的边界随流体一起运动，系统的体积、边界面的形状和大小可以随时间变化。系统的边界处没有质量交换，即没有流体流进或流出系统的边界。系统边界受到外界作用在系统上的表面力。在系统的边界上可以有能量交换，即可以有能量输入或输出系统的边界。使用系统来研究流体运动意味着采用拉格朗日的观点。在流体力学中，除个别情况外，一般采用欧拉法研究流体运动，与此相应，引入了控制体的概念。

相对于某个坐标系来说，有流体流过的固定不变的任何空间的体积称为控制体，控制体的边界面称为控制面，控制面是封闭的表面。控制面的特点如下：
① 控制面相对于坐标系是固定的；
② 在控制面上可以有质量交换，即可以有流体流进或流出控制面；
③ 控制面受到控制体以外的物体施加在控制体内流体上的力；
④ 在控制面上可以有能量交换，即可以有能量输入或输出控制面。

4.5 Continuity Equation

4.5.1 Continuity Differential Equation

An infinitesimal hexahedron in fluid field is shown in Figure 4-8. The center point of the hexahedron is $O(x, y, z)$.

The velocities of the centers of surface $ABCD$ and $EFGH$ are $u_x + \frac{\partial u_x}{\partial x} \times \frac{dx}{2}$ and $u_x - \frac{\partial u_x}{\partial x} \times \frac{dx}{2}$, respectively.

The densities of the centers of surface $ABCD$ and $EFGH$ are $\rho + \frac{\partial \rho}{\partial x} \times \frac{dx}{2}$ and $\rho - \frac{\partial \rho}{\partial x} \times \frac{dx}{2}$, respectively.

The mass from surface $EFGH$ flowing into the hexahedron in dt time is

$$\left(\rho - \frac{\partial \rho}{\partial x} \times \frac{dx}{2}\right)\left(u_x - \frac{\partial u_x}{\partial x} \times \frac{dx}{2}\right) dy\, dz\, dt$$

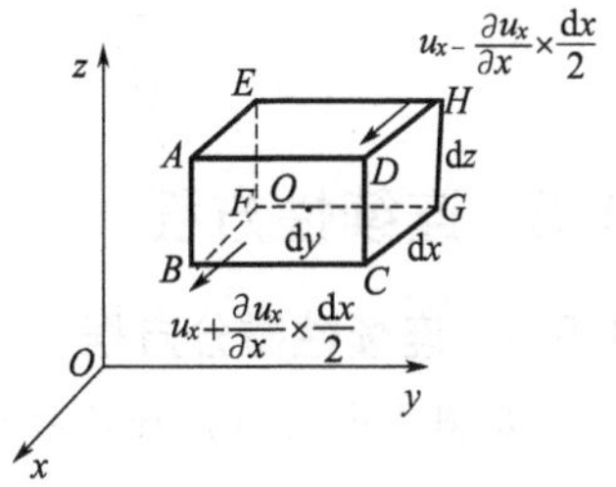

Figure 4-8 Continuity Differential Equation
图 4-8 连续性微分方程

The mass from surface ABCD flowing out the hexahedron in dt time is

$$\left(\rho + \frac{\partial \rho}{\partial x} \times \frac{dx}{2}\right)\left(u_x + \frac{\partial u_x}{\partial x} \times \frac{dx}{2}\right) dy\, dz\, dt$$

In the x direction, the mass difference between flowing into and out is $-\frac{\partial(\rho u_x)}{\partial x} dx\, dy\, dz\, dt$.

Similarly, in the y, z directions the mass differences are $-\frac{\partial(\rho u_y)}{\partial y} dx\, dy\, dz\, dt$ and $-\frac{\partial(\rho u_z)}{\partial z} dx\, dy\, dz\, dt$, respectively.

Assume that the mean density in the hexahedron is ρ. After dt time, the mean density in the hexahedron is $\rho + \dfrac{\partial \rho}{\partial t} dt$. The increment of mass in the hexahedron for density change is $\dfrac{\partial \rho}{\partial t} dx\,dy\,dz\,dt$.

According to conservation of mass, the mass difference between flowing into and out equals to the increment of mass for density change. Namely

$$\frac{\partial \rho}{\partial t} dx\,dy\,dz\,dt = -\left[\frac{\partial(\rho u_x)}{\partial x} + \frac{\partial(\rho u_y)}{\partial y} + \frac{\partial(\rho u_z)}{\partial z}\right] dx\,dy\,dz\,dt$$

$$\frac{\partial \rho}{\partial t} + \frac{\partial(\rho u_x)}{\partial x} + \frac{\partial(\rho u_y)}{\partial y} + \frac{\partial(\rho u_z)}{\partial z} = 0 \tag{4-9}$$

① For steady fluid: $\dfrac{\partial \rho}{\partial t} = 0$, and then Equation (4-9) turns into

$$\frac{\partial(\rho u_x)}{\partial x} + \frac{\partial(\rho u_y)}{\partial y} + \frac{\partial(\rho u_z)}{\partial z} = 0 \tag{4-10}$$

② For incompressible fluid: ρ is constant, so Equation (4-10) turns into

$$\frac{\partial u_x}{\partial x} + \frac{\partial u_y}{\partial y} + \frac{\partial u_z}{\partial z} = 0 \tag{4-11}$$

In column coordinate system, continuity equation is

$$\frac{\partial(\rho u_r)}{\partial r} + \frac{\partial(\rho u_\theta)}{r\,\partial \theta} + \frac{\partial(\rho u_z)}{\partial z} + \frac{\rho u_r}{r} + \frac{\partial \rho}{\partial t} = 0 \tag{4-12}$$

where u_r, u_θ, u_z——components of velocity u in r, θ and z coordinates, respectively.

In sphere coordinate system, continuity equation is

$$\frac{\partial \rho}{\partial t} + \frac{2\rho u_r}{r} + \frac{\partial \rho u_r}{\partial r} + \frac{\rho u_\theta \cot\theta}{r} + \frac{\partial(\rho u_\theta)}{r\,\partial \theta} + \frac{\partial(\rho u_\varphi)}{r\sin\theta\,\partial \varphi} = 0 \tag{4-13}$$

4.5 连续性方程

4.5.1 连续性微分方程

如图 4-8 所示，在流场中任取微元六面体，六面体的中心点为 $O(x,y,z)$。

面 $ABCD$ 和 $EFGH$ 中心点的速度分别为 $u_x + \dfrac{\partial u_x}{\partial x}\dfrac{dx}{2}$ 和 $u_x - \dfrac{\partial u_x}{\partial x}\dfrac{dx}{2}$。

面 $ABCD$ 和 $EFGH$ 中心点的密度分别为 $\rho + \dfrac{\partial \rho}{\partial x}\dfrac{dx}{2}$ 和 $\rho - \dfrac{\partial \rho}{\partial x}\dfrac{dx}{2}$。

在 dt 时间内从 $EFGH$ 面流入六面体的质量为

$$\left(\rho - \frac{\partial \rho}{\partial x}\frac{dx}{2}\right)\left(u_x - \frac{\partial u_x}{\partial x}\frac{dx}{2}\right) dy\,dz\,dt$$

在 dt 时间内从 $ABCD$ 面流出六面体的质量为

$$\left(\rho + \frac{\partial \rho}{\partial x}\frac{dx}{2}\right)\left(u_x + \frac{\partial u_x}{\partial x}\frac{dx}{2}\right) dy\,dz\,dt$$

在 x 方向，流入的质量与流出的质量之差为 $-\dfrac{\partial(\rho u_x)}{\partial x} dx\,dy\,dz\,dt$。

同理，在 y、z 方向的质量差分别为 $-\dfrac{\partial(\rho u_y)}{\partial y} dx\,dy\,dz\,dt$ 和 $-\dfrac{\partial(\rho u_z)}{\partial z} dx\,dy\,dz\,dt$。

假设六面体内的平均密度为 ρ，经过 dt 时间段后，六面体内的平均密度为 $\rho+\frac{\partial \rho}{\partial t}dt$，因密度变化引起的六面体内的质量增量为 $\frac{\partial \rho}{\partial t}dx\,dy\,dz\,dt$。

根据质量守恒定律，流入与流出的质量差等于六面体内的质量增量。即

$$\frac{\partial \rho}{\partial t}dx\,dy\,dz\,dt = -\left[\frac{\partial(\rho u_x)}{\partial x}+\frac{\partial(\rho u_y)}{\partial y}+\frac{\partial(\rho u_z)}{\partial z}\right]dx\,dy\,dz\,dt$$

$$\frac{\partial \rho}{\partial t}+\frac{\partial(\rho u_x)}{\partial x}+\frac{\partial(\rho u_y)}{\partial y}+\frac{\partial(\rho u_z)}{\partial z}=0 \tag{4-9}$$

① 恒定流体，$\frac{\partial \rho}{\partial t}=0$，则式(4-9)变为

$$\frac{\partial(\rho u_x)}{\partial x}+\frac{\partial(\rho u_y)}{\partial y}+\frac{\partial(\rho u_z)}{\partial z}=0 \tag{4-10}$$

② 不可压缩性流体，ρ 为常数，则式(4-10)变为

$$\frac{\partial u_x}{\partial x}+\frac{\partial u_y}{\partial y}+\frac{\partial u_z}{\partial z}=0 \tag{4-11}$$

在柱坐标系中，连续性方程为

$$\frac{\partial(\rho u_r)}{\partial r}+\frac{\partial(\rho u_\theta)}{r\,\partial \theta}+\frac{\partial(\rho u_z)}{\partial z}+\frac{\rho u_r}{r}+\frac{\partial \rho}{\partial t}=0 \tag{4-12}$$

式中，u_r、u_θ、u_z 是速度 u 在 r、θ、z 坐标上的分量。

在球坐标系中，连续性方程为

$$\frac{\partial \rho}{\partial t}+\frac{2\rho u_r}{r}+\frac{\partial \rho u_r}{\partial r}+\frac{\rho u_\theta \cot\theta}{r}+\frac{\partial(\rho u_\theta)}{r\,\partial \theta}+\frac{\partial(\rho u_\varphi)}{r\sin\theta\,\partial \varphi}=0 \tag{4-13}$$

4.5.2 Continuity Equation of Total Flow

The principle of conservation of mass can be applied to a flowing fluid. In Figure 4-9, there is a control volume with a fixed region in the fluid.

The masses of fluid importing and exporting control volume from dA_1 and dA_2 in dt time are $\rho_1 u_1 dA_1 dt$, and $\rho_2 u_2 dA_2 dt$, respectively.

For steady flow, the mass of fluid in the control volume remains constant, and the relation is

$$\rho_1 u_1 dA_1 = \rho_2 u_2 dA_2 \tag{4-14}$$

For an incompressible fluid, $\rho_1 = \rho_2$, Equation (4-14) turns into

$$u_1 dA_1 = u_2 dA_2 \tag{4-15}$$

Because $u\,dA = dQ$, Equation (4-15) can be written as

$$dQ = u_1 dA_1 = u_2 dA_2 \tag{4-16}$$

Integrating

$$\int dQ = \int_{A_1} u_1 dA_1 = \int_{A_2} u_2 dA_2$$

$$Q = v_1 A_1 = v_2 A_2 \tag{4-17}$$

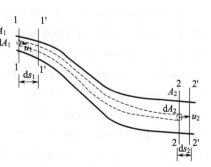

Figure 4-9 The Derivation of Continuity Equation for Total Flow

图 4-9 总流的连续性方程的推导

This is the continuity equation of total flow. The continuity equation can also be applied to determining

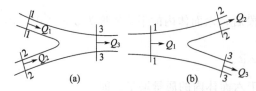

Figure 4-10 Conflux (a) and Distributary (b)

图 4-10 汇流 (a) 与分流 (b)

the relation between the flows into and out of a junction. In Figure 4-10, the total mass flowing into the junction equals to the total mass flowing out of the junction.

In Figure 4-10(a)

$$Q_1+Q_2=Q_3, \quad v_1A_1+v_2A_2=v_3A_3 \tag{4-18}$$

In Figure 4-10(b)

$$Q_1=Q_2+Q_3, \quad v_1A_1=v_2A_2+v_3A_3 \tag{4-19}$$

4.5.2 总流的连续性方程

质量守恒定律可应用于流体流动。在图 4-9 中，有一确定边界的控制体。

在 dt 时间内，从 dA_1 流入的质量为 $\rho_1 u_1 dA_1 dt$，从 dA_2 流出的质量为 $\rho_2 u_2 dA_2 dt$。对于恒定流，控制体内的质量是常数，则以上两式的关系为

$$\rho_1 u_1 dA_1 = \rho_2 u_2 dA_2 \tag{4-14}$$

对于不可压缩均质流体，$\rho_1=\rho_2$，式(4-14) 转化为：

$$u_1 dA_1 = u_2 dA_2 \tag{4-15}$$

因 $u dA = dQ$，所以式(4-15) 也可写作

$$dQ = u_1 dA_1 = u_2 dA_2 \tag{4-16}$$

积分得到

$$\int dQ = \int_{A_1} u_1 dA_1 = \int_{A_2} u_2 dA_2$$

$$Q = v_1 A_1 = v_2 A_2 \tag{4-17}$$

这就是总流的连续性方程。若沿途有流量流进或流出，总流的连续性方程仍然适用，只是形式有所不同。如图 4-10 所示，流入结点的流量等于流出结点的流量。

在图 4-10(a) 中

$$Q_1+Q_2=Q_3, \quad v_1A_1+v_2A_2=v_3A_3 \tag{4-18}$$

在图 4-10(b) 中

$$Q_1=Q_2+Q_3, \quad v_1A_1=v_2A_2+v_3A_3 \tag{4-19}$$

4.6 Motion Differential Equation of Ideal Fluid

In the previous section the continuity equation was discussed. It reflects the conditions which velocity field of fluid motion must satisfy. It is a kinematics equation. Now let us analyze the kinematics relations between the stress and fluid motion of fluid. That is to build the kinematics equation of ideal fluid.

Consider the infinitesimal right-angled hexahedron whose length of sides are dx, dy, dz, as shown in Figure 4-11. In it the coordinate of center point M is $M(x, y, z)$, and its pressure is p. There are two kinds of outside forces acting on this right-angled hexahedron: surface force and body force.

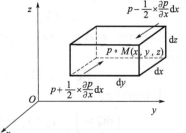

Figure 4-11 Euler's Motion Differential Equation

图 4-11 欧拉运动微分方程

Suppose the unit body forces in the x, y and z direction are X, Y, Z. The density of the fluid is ρ, and three components of acceleration are $\dfrac{du_x}{dt}$, $\dfrac{du_y}{dt}$, $\dfrac{du_z}{dt}$.

According to Newton's second law, the motion equation in x direction is

$$\left(p - \frac{1}{2} \times \frac{\partial p}{\partial x} dx\right) dy\,dz - \left(p + \frac{1}{2} \times \frac{\partial p}{\partial x} dx\right) dy\,dz + f_x \rho\, dx\,dy\,dz = \rho\, dx\,dy\,dz\, \frac{du_x}{dt}$$

In a similar way, the motion equations in y and z direction are

$$\left(p - \frac{1}{2} \times \frac{\partial p}{\partial y} dy\right) dx\,dz - \left(p + \frac{1}{2} \times \frac{\partial p}{\partial y} dy\right) dx\,dz + f_y \rho\, dx\,dy\,dz = \rho\, dx\,dy\,dz\, \frac{du_y}{dt}$$

$$\left(p - \frac{1}{2} \times \frac{\partial p}{\partial z} dz\right) dx\,dy - \left(p + \frac{1}{2} \times \frac{\partial p}{\partial z} dz\right) dx\,dy + f_z \rho\, dx\,dy\,dz = \rho\, dx\,dy\,dz\, \frac{du_z}{dt}$$

After simplifying the above equations, the result is

$$\left.\begin{aligned}\frac{du_x}{dt} &= f_x - \frac{1}{\rho} \times \frac{\partial p}{\partial x} \\ \frac{du_y}{dt} &= f_y - \frac{1}{\rho} \times \frac{\partial p}{\partial y} \\ \frac{du_z}{dt} &= f_z - \frac{1}{\rho} \times \frac{\partial p}{\partial z}\end{aligned}\right\} \qquad (4\text{-}20)$$

Substituting Equation(4-5) into Equation(4-20) yields

$$\left.\begin{aligned}\frac{\partial u_x}{\partial t} + u_x \frac{\partial u_x}{\partial x} + u_y \frac{\partial u_x}{\partial y} + u_z \frac{\partial u_x}{\partial z} &= f_x - \frac{1}{\rho} \times \frac{\partial p}{\partial x} \\ \frac{\partial u_y}{\partial t} + u_x \frac{\partial u_y}{\partial x} + u_y \frac{\partial u_y}{\partial y} + u_z \frac{\partial u_y}{\partial z} &= f_y - \frac{1}{\rho} \times \frac{\partial p}{\partial y} \\ \frac{\partial u_z}{\partial t} + u_x \frac{\partial u_z}{\partial x} + u_y \frac{\partial u_z}{\partial y} + u_z \frac{\partial u_z}{\partial z} &= f_z - \frac{1}{\rho} \times \frac{\partial p}{\partial z}\end{aligned}\right\} \qquad (4\text{-}21)$$

The above two equations are motion differential equations of ideal fluid. They are also called Euler's motion differential equation.

In these equations, x, y, z and t are four variables; p, u_x, u_y and u_z, functions of x, y, z and t, are unknown quantities; f_x, f_y, f_z are also functions of x, y and z, but they are known normally.

4.6 理想流体的运动微分方程

上节讨论了连续性方程，它反映了流体运动速度场必须满足的条件，这是一个运动学方程。现在我们分析流体受力及运动之间的动力学关系，即建立理想流体运动学方程。

考虑如图 4-11 所示的边长为 dx、dy、dz 的微元直角六面体，中点 M 坐标为 $M(x, y, z)$，该点的压强为 p。作用在此直角六面体上的外力有两种：表面压力和质量力。

设在 x、y、z 轴方向上的单位质量力为 X、Y、Z，又设流体的密度为 ρ，加速度的三个分量为 $\dfrac{du_x}{dt}$、$\dfrac{du_y}{dt}$、$\dfrac{du_z}{dt}$。

根据牛顿第二定律得 x 方向的运动方程式为

$$\left(p - \frac{1}{2} \times \frac{\partial p}{\partial x} dx\right) dy\,dz - \left(p + \frac{1}{2} \times \frac{\partial p}{\partial x} dx\right) dy\,dz + f_x \rho\, dx\,dy\,dz = \rho\, dx\,dy\,dz\, \frac{du_x}{dt}$$

同理，y、z 轴方向上的运动方程式为

$$\left(p - \frac{1}{2} \times \frac{\partial p}{\partial y} dy\right) dx dz - \left(p + \frac{1}{2} \times \frac{\partial p}{\partial y} dy\right) dx dz + f_y \rho dx dy dz = \rho dx dy dz \frac{du_y}{dt}$$

$$\left(p - \frac{1}{2} \times \frac{\partial p}{\partial z} dz\right) dx dy - \left(p + \frac{1}{2} \times \frac{\partial p}{\partial z} dz\right) dx dy + f_z \rho dx dy dz = \rho dx dy dz \frac{du_z}{dt}$$

上式简化后得

$$\left. \begin{aligned} \frac{du_x}{dt} &= f_x - \frac{1}{\rho} \times \frac{\partial p}{\partial x} \\ \frac{du_y}{dt} &= f_y - \frac{1}{\rho} \times \frac{\partial p}{\partial y} \\ \frac{du_z}{dt} &= f_z - \frac{1}{\rho} \times \frac{\partial p}{\partial z} \end{aligned} \right\} \quad (4\text{-}20)$$

将式(4-5)代入式(4-20)得

$$\left. \begin{aligned} \frac{\partial u_x}{\partial t} + u_x \frac{\partial u_x}{\partial x} + u_y \frac{\partial u_x}{\partial y} + u_z \frac{\partial u_x}{\partial z} &= f_x - \frac{1}{\rho} \times \frac{\partial p}{\partial x} \\ \frac{\partial u_y}{\partial t} + u_x \frac{\partial u_y}{\partial x} + u_y \frac{\partial u_y}{\partial y} + u_z \frac{\partial u_y}{\partial z} &= f_y - \frac{1}{\rho} \times \frac{\partial p}{\partial y} \\ \frac{\partial u_z}{\partial t} + u_x \frac{\partial u_z}{\partial x} + u_y \frac{\partial u_z}{\partial y} + u_z \frac{\partial u_z}{\partial z} &= f_z - \frac{1}{\rho} \times \frac{\partial p}{\partial z} \end{aligned} \right\} \quad (4\text{-}21)$$

上面二式即是理想流体运动的微分方程式，也叫作欧拉运动微分方程式。

式中 x、y、z、t 为四个变量；ρ、u_x、u_y、u_z 为 x、y、z、t 的函数，是未知量；f_x、f_y、f_z 也是 x、y、z 的函数，一般是已知的。

4.7 Bernoulli's Equation and Its Application

Bernoulli's equation is the embodiment of the law of conservation and translation of energy in fluid mechanics.

4.7.1 Bernoulli's Equation for Element Flow of an Ideal Fluid

Because there are so many basic assumptions involved in derivation of Bernoulli's equation, it is important to remember them all when applying it. The basic assumptions are as follows.

① The fluid is inviscid;
② The fluid is incompressible;
③ The flow is steady;
④ The force has potential;
⑤ The flow is irrotational;
⑥ No energy is added to or removed from the fluid along the streamline.

As shown in Figure 4-12, let z_1, u_1, dA_1, ds_1 and z_2, u_2, dA_2, ds_2 be the elevation, velocity, area and distance at sections 1 and 2, respectively. Applying the principle of kinetic energy, the increment of kinetic energy is equal to the work of all outside forces.

$$\Delta E = W$$

When the element flow moves from 1-2 to $1'$-$2'$, the increment of kinetic energy between 1-2 and $1'$-$2'$ is ΔE.

$$\frac{1}{2}mu_2^2 - \frac{1}{2}mu_1^2 = \frac{\rho \mathrm{d}s_2 \mathrm{d}A_2 u_2^2}{2} - \frac{\rho \mathrm{d}s_1 \mathrm{d}A_1 u_1^2}{2}$$

$$= \frac{\gamma \mathrm{d}Q \mathrm{d}t u_2^2}{2g} - \frac{\gamma \mathrm{d}Q \mathrm{d}t u_1^2}{2g}$$

$$= \frac{\gamma \mathrm{d}Q \mathrm{d}t}{g}\left(\frac{u_2^2}{2} - \frac{u_1^2}{2}\right)$$

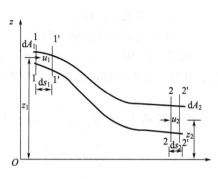

Figure 4-12 The Derivation of Bernoulli's Equation for Element Flow of Ideal Fluid

图 4-12 理想流体元流伯努利方程推导

Outside forces include gravity and pressure force. Work done by gravity and pressure force are as follows.

Work done by gravity

$$W_1 = \gamma \mathrm{d}A_1 \mathrm{d}s_1 (z_1 - z_2) = \gamma \mathrm{d}Q \mathrm{d}t (z_1 - z_2)$$

A steadily flowing stream of fluid can do work because of pressure. At any given cross-section, the pressure generates a force. Work will be done accompanying the movement of cross-section.

Work done by pressure

$$W_2 = p_1 \mathrm{d}A_1 u_1 \mathrm{d}t - p_2 \mathrm{d}A_2 u_2 \mathrm{d}t = \mathrm{d}Q \mathrm{d}t (p_1 - p_2)$$

According to the principle of kinetic energy, the kinetic energy increment equals to the work of outside forces.

The increment of kinetic energy

$$\Delta E = \text{Work done by gravity } W_1 + \text{Work done by pressure } W_2$$

$$\frac{\gamma \mathrm{d}Q \mathrm{d}t}{g}\left(\frac{u_2^2}{2} - \frac{u_1^2}{2}\right) = \gamma \mathrm{d}Q \mathrm{d}t \ (z_1 - z_2) + \mathrm{d}Q \mathrm{d}t \ (p_1 - p_2) \qquad [4\text{-}22(a)]$$

$$z + \frac{p}{\gamma} + \frac{u^2}{2g} = \text{constant}$$

Equation [4-22(a)] is Bernoulli's equation for an incompressible ideal fluid along streamlines under the steady flow conditions.

Bernoulli's equation for element flow of real fluid is

$$z_1 + \frac{p_1}{\gamma} + \frac{u_1^2}{2g} = z_2 + \frac{p_2}{\gamma} + \frac{u_2^2}{2g} + h_w' \qquad [4\text{-}22(b)]$$

4.7 伯努利方程及其应用

伯努利方程是能量守恒与转换定律在流体力学中的具体体现。

4.7.1 理想流体元流的伯努利方程

由于伯努利方程推导中包含了一些基本假设，应用该方程时应该知道这些假设。这些基本假设如下：

① 流体无黏性；

② 流体不可压缩；

③ 流体是恒定流；

④ 作用力是有势的力；
⑤ 流体是无旋运动；
⑥ 沿流线无能量的输入与输出。

如图 4-12 所示，z_1、u_1、dA_1、ds_1 和 z_2、u_2、dA_2、ds_2 分别是断面 1 和 2 的位置高度、速度、面积及移动的距离。应用动能定理，动能的增量等于外力做的功。

$$\Delta E = W$$

当元流从 1—2 运动到 $1'$—$2'$，从 1—2 到 $1'$—$2'$ 的动能的增量为 ΔE。

$$\frac{1}{2}mu_2^2 - \frac{1}{2}mu_1^2 = \frac{\rho ds_2 dA_2 u_2^2}{2} - \frac{\rho ds_1 dA_1 u_1^2}{2}$$

$$= \frac{\gamma dQ dt u_2^2}{2g} - \frac{\gamma dQ dt u_1^2}{2g} = \frac{\gamma dQ dt}{g}\left(\frac{u_2^2}{2} - \frac{u_1^2}{2}\right)$$

外力包括重力与压力两种力，重力和压力做的功分别为：

重力做功 $\qquad W_1 = \gamma dA_1 ds_1(z_1 - z_2) = \gamma dQ dt(z_1 - z_2)$

恒定流体中压力可以做功，在任意断面上，压力将产生一种力，并随断面运动而做功。

压力做功 $\qquad W_2 = p_1 dA_1 u_1 dt - p_2 dA_2 u_2 dt = dQ dt(p_1 - p_2)$

根据动能定理，动能的增量等于外力做功。

动能的增量 $\Delta E =$ 重力做功 $W_1 +$ 压力做功 W_2

$$\frac{\gamma dQ dt}{g}\left(\frac{u_2^2}{2} - \frac{u_1^2}{2}\right) = \gamma dQ dt\ (z_1 - z_2)\ + dQ dt\ (p_1 - p_2) \qquad [4\text{-}22(a)]$$

$$z + \frac{p}{\gamma} + \frac{u^2}{2g} = 常数$$

式[4-22(a)]就是不可压缩理想流体在稳定流条件下沿流线的伯努利方程。

实际流体的元流伯努利方程为

$$z_1 + \frac{p_1}{\gamma} + \frac{u_1^2}{2g} = z_2 + \frac{p_2}{\gamma} + \frac{u_2^2}{2g} + h'_w \qquad [4\text{-}22(b)]$$

(1) Physical meaning

$\qquad z$—potential energy of unit weight fluid, m;

$\qquad p/\gamma$—pressure energy of unit weight fluid, m;

$\qquad u^2/(2g)$—kinetic energy of unit weight fluid, m;

$\qquad h'_w$—energy loss of unit weight fluid, m;

$\qquad z + p/\gamma$—potential energy of unit weight fluid, m;

$z + p/\gamma + u^2/(2g)$—mechanical energy of unit weight fluid, m.

(1) 物理意义 z 表示单位重量流体的势能，m；p/γ 表示单位重量流体的压力能，m；$u^2/(2g)$ 表示单位重量流体的动能，m；h'_w 表示单位重量流体的能量损失，m；$z + p/\gamma$ 表示单位重量流体的势能，m；$z + p/\gamma + u^2/(2g)$ 表示单位重量流体的总机械能，m。

(2) Geometric meaning Geometric meaning of Bernoulli's equation for element flow is expressed in Figure 4-13.

$\qquad z$—the height from the position of a certain point to the datum plane which is called position head, m;

p/γ— the height of liquid column acted by point pressure which is called pressure head, m;

$u^2/(2g)$—velocity head, m;

h'_w—head loss, m;

$z+p/\gamma$—piezometer head, m;

$z+p/\gamma+u^2/(2g)$—total head, m.

（2）几何意义　图 4-13 表示元流伯努利方程的几何意义。z 表示某点位置到基准面的高度，称为位置水头，m；p/γ 表示某点压强作用下液柱的高度，称为压强水头，m；$u^2/(2g)$ 表示流速水头，m；h'_w 表示水头损失，m；$z+p/\gamma$ 表示测压管水头，m；$z+p/\gamma+u^2/(2g)$ 为总水头，m。

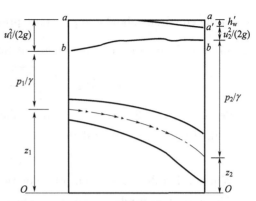

Figure 4-13　Geometric Meaning of Bernoulli's Equation for Element Flow

图 4-13　元流伯努利方程的几何意义

4.7.2　Application of Bernoulli's Equation of Element Flow for the Ideal Fluids—Pitot Tube

Pitot tube is an apparatus which transforms the kinetic energy of fluid into pressure energy and then uses manometer to measure the motion velocity of fluid. It is usually used to measure the velocity of flow in channel, open channel and air conduit and also to measure the object motion velocity in fluid such as ships and airplanes etc. Pitot tubes can be simple or composite. Their structures and measurement principles are shown in Figure 4-14 and Figure 4-15.

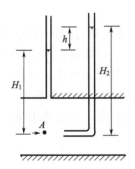

Figure 4-14　Simple Pitot Tube

图 4-14　简易皮托管

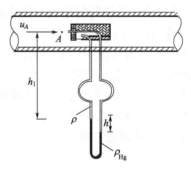

Figure 4-15　Composite Pitot Tube

图 4-15　复合皮托管

The simple Pitot tube was designed based on the principle that the surface of liquid in the tube ascends because the velocity of stagnation point is zero and the kinetic energy transforms into the pressure energy. The point of zero velocity is called stagnation point and the pressure of this point is named stagnation pressure p_s. According to the energy equation, we can obtain the following equation.

$$\frac{p_A}{\gamma}+\frac{u_A^2}{2g}=\frac{p_s}{\gamma}$$

$$p_s=p_A+\frac{1}{2}\rho u_A^2$$

In the equations above, p_A is static pressure, $\frac{1}{2}\rho u_A^2$ is dynamic pressure and p_s is total pressure. Its theory formula of velocity of flow is as follows.

$$v=\sqrt{2g\frac{(p_s-p_A)}{\gamma}}=\sqrt{2\frac{(p_s-p_A)}{\rho}}=\sqrt{2gh} \qquad (4\text{-}23)$$

Because of the disturbance of Pitot tube structure to the fluid field, the formula should be modified in the accurate calculation.

$$v=c\sqrt{2gh}$$

Where c is called coefficient of velocity of flow, normally $c=0.98\sim 0.995$.

There are two kinds of instances when using composite Pitot tubes to measure velocity of fluid.

(1) For liquid in the tube

$$v=\sqrt{2g\frac{(\gamma'-\gamma)}{\gamma}h}=\sqrt{2g\frac{(\rho'-\rho)}{\rho}h}$$

(2) For air in the tube

$$\rho'\gg\rho,\quad \rho'-\rho\approx\rho'$$

$$v=\sqrt{2g\frac{\rho'}{\rho}h}$$

The Pitot tubes used in engineering must be strictly calibrated, and the measurement conditions and the type of fluid must be described. Moreover, in order to lessen the measure error, installation should be done according to the specification.

4.7.2 理想流体元流的伯努利方程的应用——皮托管

皮托管（Pitot tube）是指将流体动能转化为压力能，进而通过测压计测定流体运动速度的仪器。皮托管常用于测量河道、明渠、风管中的流速，还可测量物体在流体中的运动速度，如测量船舶、飞机等的航行速度可用皮托管。皮托管有简单和复合之分，其结构及测量原理如图 4-14、图 4-15 所示。简易皮托管是依据驻点流速为零，其动能转变为压力能，从而使管内液面上升的原理设计成的。速度为零处的端点称为驻点，该点的压强称驻点压强或滞止压强 p_s，由能量方程可得

$$\frac{p_A}{\gamma}+\frac{u_A^2}{2g}=\frac{p_s}{\gamma}$$

$$p_s=p_A+\frac{1}{2}\rho u_A^2$$

式中，p_A 为静压；$\frac{1}{2}\rho u_A^2$ 为动压；p_s 为全压或总压。其理论流速公式为：

$$v=\sqrt{2g\frac{(p_s-p_A)}{\gamma}}=\sqrt{2\frac{(p_s-p_A)}{\rho}}=\sqrt{2gh} \qquad (4\text{-}23)$$

由于皮托管结构对流场的扰动，在精确计算时应对公式进行修正：

$$v=c\sqrt{2gh}$$

式中，c 称为流速系数，一般 $c=0.98\sim 0.995$。

用复合皮托管测量流速时，有以下两种情况。

(1) 管中为液体

$$v=\sqrt{2g\frac{(\gamma'-\gamma)}{\gamma}h}=\sqrt{2g\frac{(\rho'-\rho)}{\rho}h}$$

(2) 管中为气体

$$\rho'\gg\rho,\ \rho'-\rho\approx\rho'$$

$$v=\sqrt{2g\frac{\rho'}{\rho}h}$$

工程中使用的皮托管都必须经过严格标定，说明测量条件和流体种类，而且在安装时应按说明书要求去做，以减少测量误差。

4.7.3 Bernoulli's Equation of Total Flow

In engineering practice we often need to solve the total flow problems, such as the problems that fluids flow in pipes or channels. So we need extend Bernoulli's equation to the total flow by integrating on cross-section of flow additionally.

Multiply each item of Equation[4-22 (b)] by $\rho g\, dQ$, and then the mechanical energy relationship of the total flow across two cross-sections of flow of infinitesimal streamline tube in unit time is

$$\left(z_1+\frac{p_1}{\gamma}+\frac{u_1^2}{2g}\right)\gamma u_1 dA_1 = \left(z_2+\frac{p_2}{\gamma}+\frac{u_2^2}{2g}\right)\gamma u_2 dA_2 + h'_w \gamma dQ$$

where $dQ = u_1 dA_1 = u_2 dA_2$. After integrating we obtain that the whole mechanical energy relationship between two cross-sections of total flow is

$$\int_{A_1}\left(z_1+\frac{p_1}{\gamma}+\frac{u_1^2}{2g}\right)\gamma u_1 dA_1 = \int_{A_2}\left(z_2+\frac{p_2}{\gamma}+\frac{u_2^2}{2g}\right)\gamma u_2 dA_2 + \int_Q h'_w \gamma dQ \qquad (4\text{-}24)$$

in which

(1) $\int_A \left(z+\frac{p}{\gamma}\right)\gamma u\, dA$　It is the summation of potential energy and pressure energy which goes through the cross-section of fluid in unit time.

At the suddenly-varying section, the $(z+\frac{p}{\gamma})$ of each point is not constant and it's difficult to integrate.

At the gradually-varying section, the distribution of the dynamic pressure accords to the static pressure approximately, the $(z+\frac{p}{\gamma})$ of each point is constant.

So if choose a gradually-varying cross-section, and then integrate, we obtain

$$\int_A \left(z+\frac{p}{\gamma}\right)\gamma u\, dA = \left(z+\frac{p}{\gamma}\right)\gamma \int_A u\, dA$$

$$\int_A u\, dA = Q$$

$$\int_A \left(z+\frac{p}{\gamma}\right)\gamma u\, dA = \left(z+\frac{p}{\gamma}\right)\gamma Q \qquad (4\text{-}25)$$

(2) $\int_A \frac{u^2}{2g}\gamma u\, dA$　It is the summation of fluid kinetic energy which goes through the

cross-section of total flow in unit time.

$$\int_A \frac{u^2}{2g}\gamma u\,dA = \frac{\gamma}{2g}\int_A u^3\,dA$$

where
$$\int_A u^3\,dA = \int_A (v+\Delta u)^3\,dA = \int_A (v^3 + 3v^2\Delta u + 3v\Delta u^2 + \Delta u^3)\,dA$$
$$= v^3 A + 3v^2 \int_A \Delta u\,dA + 3v\int_A \Delta u^2\,dA + \int_A \Delta u^3\,dA$$

For $\int_A \Delta u\,dA = 0$, $\int_A \Delta u^3\,dA$ can be neglected.

We can obtain
$$\int_A u^3\,dA = v^3 A + 3v\int_A \Delta u^2\,dA = \alpha v^3 A$$

or
$$\alpha = \frac{\int_A u^3\,dA}{v^3 A} = \frac{\int_A u^3\,dA}{v^2 Q}$$

In the equations above, α is the kinetic-energy correction factor. The usual value of α is 1.05~1.10. In engineering calculation, $\alpha = 1.0$ is often used.

Because the velocity distribution on cross-section is difficult to confirm, the average velocity v is often used to denote the factual kinetic energy in engineering, namely

$$\int_A \frac{u^2}{2g}\gamma u\,dA = \gamma \frac{\alpha v^3}{2g}A = \alpha \frac{v^2}{2g}\gamma Q \tag{4-26}$$

4.7.3 总流伯努利方程

在工程实际中我们往往要解决总流流动问题，如流体在管道、渠道中的流动问题，因此还需要通过在过流断面上积分把伯努利方程推广到总流上去。

将式[4-22（b）]各项同乘 $\rho g\,dQ$，则单位时间内通过微元流束两个过流断面的全部流体的机械能的关系式为

$$\left(z_1 + \frac{p_1}{\gamma} + \frac{u_1^2}{2g}\right)\gamma u_1\,dA_1 = \left(z_2 + \frac{p_2}{\gamma} + \frac{u_2^2}{2g}\right)\gamma u_2\,dA_2 + h'_w \gamma\,dQ$$

其中 $dQ = u_1\,dA_1 = u_2\,dA_2$。积分得通过总流两个过流断面的总机械能之间的关系式为

$$\int_{A_1}\left(z_1 + \frac{p_1}{\gamma} + \frac{u_1^2}{2g}\right)\gamma u_1\,dA_1 = \int_{A_2}\left(z_2 + \frac{p_2}{\gamma} + \frac{u_2^2}{2g}\right)\gamma u_2\,dA_2 + \int_Q h'_w \gamma\,dQ \tag{4-24}$$

其中：

（1）$\int_A \left(z + \frac{p}{\gamma}\right)\gamma u\,dA$　它是单位时间内通过总流过流断面的流体势能和压力能的总和。

在急变流断面上，各点的 $\left(z + \frac{p}{\gamma}\right)$ 不为常数，积分困难。在渐变流断面上，流体动压强近似地按静压强分布，各点的 $\left(z + \frac{p}{\gamma}\right)$ 为常数。

因此，若将过流断面取在渐变流断面上，积分后可得

$$\int_A \left(z + \frac{p}{\gamma}\right)\gamma u\,dA = \left(z + \frac{p}{\gamma}\right)\gamma \int_A u\,dA$$

$$\int_A u \, dA = Q$$

$$\int_A \left(z + \frac{p}{\gamma}\right) \gamma u \, dA = \left(z + \frac{p}{\gamma}\right) \gamma Q \tag{4-25}$$

(2) $\int_A \dfrac{u^2}{2g} \gamma u \, dA$　它是单位时间内通过总流过流断面的流体动能的总和。

$$\int_A \frac{u^2}{2g} \gamma u \, dA = \frac{\gamma}{2g} \int_A u^3 \, dA$$

其中

$$\int_A u^3 \, dA = \int_A (v + \Delta u)^3 \, dA = \int_A (v^3 + 3v^2 \Delta u + 3v \Delta u^2 + \Delta u^3) \, dA$$

$$= v^3 A + 3v^2 \int_A \Delta u \, dA + 3v \int_A \Delta u^2 \, dA + \int_A \Delta u^3 \, dA$$

因为 $\int_A \Delta u \, dA = 0$，$\int_A \Delta u^3 \, dA$ 可以忽略。

可得

$$\int_A u^3 \, dA = v^3 A + 3v \int_A \Delta u^2 \, dA = \alpha v^3 A$$

或者

$$\alpha = \frac{\int_A u^3 \, dA}{v^3 A} = \frac{\int_A u^3 \, dA}{v^2 Q}$$

式中，α 为动能修正系数，一般可取 1.05~1.10，工程计算中常取 $\alpha = 1.0$。

由于过流断面上的速度分布一般难以确定，工程上常用断面平均速度 v 来表示实际动能，即

$$\int_A \frac{u^2}{2g} \gamma u \, dA = \gamma \frac{\alpha v^3}{2g} A = \alpha \frac{v^2}{2g} \gamma Q \tag{4-26}$$

(3) $\int_Q h'_w \gamma \, dQ$　h_w is the average energy loss of unit weight fluid between sections 1-1 and 2-2 on total flow.

$$\int_Q h'_w \gamma \, dQ = h_w \gamma Q \tag{4-27}$$

Substitute Equation (4-25) and Equation (4-26) into Equation (4-24), we can obtain the following equation.

$$z_1 + \frac{p_1}{\gamma} + \frac{\alpha_1 v_1^2}{2g} = z_2 + \frac{p_2}{\gamma} + \frac{\alpha_2 v_2^2}{2g} \tag{4-28}$$

This is the Bernoulli's equation of total flow for ideal fluid.

The real fluid has viscosity. Because the internal frictional resistances between the fluid layers do work, a portion of mechanical energy is consumed and turns into heat energy. So the Bernoulli's equation of total flow for real fluid is

$$z_1 + \frac{p_1}{\gamma} + \frac{\alpha_1 v_1^2}{2g} = z_2 + \frac{p_2}{\gamma} + \frac{\alpha_2 v_2^2}{2g} + h_w \tag{4-29}$$

where　h_w——the average energy loss of unit weight fluid between section 1-1 and 2-2 on total flow.

(3) $\int_Q h'_w \gamma dQ$　令 h_w 为单位重量流体由过流断面 1—1 移动到过流断面 2—2 时能量损失的平均值，则可以得到

$$\int_Q h'_w \gamma dQ = h_w \gamma Q \tag{4-27}$$

将式(4-25)、式(4-26)代入式(4-24)，整理得到

$$z_1 + \frac{p_1}{\gamma} + \frac{\alpha_1 v_1^2}{2g} = z_2 + \frac{p_2}{\gamma} + \frac{\alpha_2 v_2^2}{2g} \tag{4-28}$$

这就是理想流体总流的伯努利方程。

实际流体有黏性，由于流层间内摩擦阻力做功会消耗部分机械能转化为热能，因此实际流体总流的伯努利方程为：

$$z_1 + \frac{p_1}{\gamma} + \frac{\alpha_1 v_1^2}{2g} = z_2 + \frac{p_2}{\gamma} + \frac{\alpha_2 v_2^2}{2g} + h_w \tag{4-29}$$

式中，h_w 为总流过流断面上 1—1 和 2—2 两个断面之间单位重量流体的平均能量损失。

4.7.4　The Physical and Geometric Meaning of Bernoulli's Equation

(1) Physical meaning

z—position energy of unit weight fluid, m;

p/γ—pressure energy of unit weight fluid, m;

$v^2/(2g)$—kinetic energy of unit weight fluid, m;

h_w—energy loss of unit weight fluid, m;

$z+p/\gamma$—potential energy of unit weight fluid, m;

$z+p/\gamma+v^2/(2g)$—mechanical energy of unit weight fluid, m.

(2) Geometric meaning

z—position head, m;

p/γ—pressure head, m;

$v^2/(2g)$—velocity head, m;

h_w—head loss, m;

$z+p/\gamma$—piezometer head, m;

$z+p/\gamma+v^2/(2g)$—total head, m.

(3) Head line. There are four kinds of lines in Figure 4-16.

① Datum line. Datum line is a horizontal line.

② Center line. Center line is a line connected by the center points of a fluid.

③ Piezometric head line or hydraulic grade line (HGL). It represents the level to which liquid will rise in a piezometer tube. It may be ascending, declining and horizontal.

④ Total energy line or total head line. Total head line is a horizontal line for ideal liquid, but it will be a declining line for real liquid.

A Pitot tube with its open in the flow pointing upstream will intercept the kinetic energy of the flow in addition to the piezometric head, and so its liquid level indicates the total energy head. The vertical distance between the liquid surface in the piezometer tube and that in the Pitot tube is $v^2/(2g)$. For ideal liquid, the total energy line is a horizontal line. For real liquid, the total energy line is declining. Hydraulic gradient can be expressed as follows.

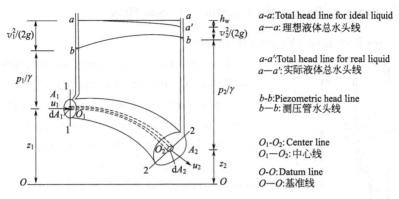

Figure 4-16　Head Line
图 4-16　水头线

$$J = \frac{dh_w}{dl} = -\frac{dh}{dl} \tag{4-30}$$

4.7.4　伯努利方程的物理意义与几何意义

（1）物理意义

z——单位重量流体的势能，m；

p/γ——单位重量流体的压力能，m；

$v^2/(2g)$——单位重量流体的动能，m；

h_w——单位重量流体的能量损失，m；

$z+p/\gamma$——单位重量流体的势能，m；

$z+p/\gamma+v^2/(2g)$——单位重量流体的总机械能，m。

（2）几何意义

z——位置水头，m；

p/γ——压强水头，m；

$v^2/(2g)$——流速水头，m；

h_w——水头损失，m；

$z+p/\gamma$——测压管水头，m；

$z+p/\gamma+v^2/(2g)$——总水头，m。

（3）水头线　图 4-16 中有四种类型的线条。

① 基准线。基准线是一条水平线。

② 中心线。中心线是流体的中心点连成的线。

③ 测压管水头线或水力坡度线。测压管水头线表示测压管内上升的液位连成的线。有上升、下降及水平三种情况。

④ 总水头线。理想流体的总水头线为水平线，实际流体的总水头线为下降的曲线。

皮托管内正对来水方向的一肢转化了流体的动能，其液位表示总水头。该液位与测压管内的液位垂直距离之差为 $v^2/(2g)$。理想流体的总水头线为水平线。实际流体的总水头线是向下倾斜的，可以用水力坡度来表达其下降的程度。

$$J = \frac{dh_w}{dl} = -\frac{dh}{dl} \tag{4-30}$$

4.7.5 Applications of Bernoulli's Equation of Total Flow

(1) Measure flow rate—Venturi flowmeter

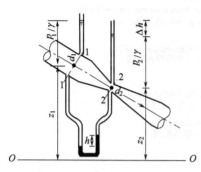

Figure 4-17　Venturi Flowmeter
图 4-17　文丘里流量计

The fluid section contracts as the liquid in the conduit flows via the throttle equipment. The increasing of velocity of flow and the falling of pressure on the contracting section bring the differential pressure on the forward and backward of the throttle equipment.

Now we use Venturi flowmeter as an example to deduce the formula of calculating the flux. Venturi flowmeter is a kind of apparatus to measure the fluid flux in the conduit under pressure. As shown in Figure 4-17, it consists of three parts: slick constricted section, throat and expansion section.

Choose section 1-1 and section 2-2, and the calculation points are all on conduit. Datum plane 0-0 is on a fixed position under the conduit and assume $\alpha_1 = \alpha_2 = 1.0$. Neglect friction, and the Bernoulli's equation of total flow between section 1-1 and section 2-2 is

$$z_1 + \frac{p_1}{\gamma} + \frac{v_1^2}{2g} = z_2 + \frac{p_2}{\gamma} + \frac{v_2^2}{2g}$$

From continuity equation, we obtain

$$v_1 A_1 = v_2 A_2$$

Combine the two formulas above to obtain

$$v_1 = \frac{1}{\sqrt{\left(\frac{A_1}{A_2}\right)^2 - 1}} \sqrt{2g\left[\left(z_1 + \frac{p_1}{\rho g}\right) - \left(z_2 + \frac{p_2}{\rho g}\right)\right]} = \frac{1}{\sqrt{\left(\frac{A_1^2}{A_2^2}\right) - 1}} \sqrt{2g\Delta h}$$

So the volume flux through the flowmeter is

$$Q' = v_1 A_1 = \frac{A_1}{\sqrt{\left(\frac{A_1}{A_2}\right)^2 - 1}} \sqrt{2g\Delta h} = \frac{\pi d_1^2 d_2^2}{4\sqrt{d_1^4 - d_2^4}} \sqrt{2g\Delta h}$$

Considering the influence of fluid viscosity, the right of the above formula should multiply a correction factor of flux. Then

$$Q = \mu \frac{\pi d_1^2 d_2^2}{4\sqrt{d_1^4 - d_2^4}} \sqrt{2g\Delta h} \tag{4-31}$$

4.7.5　总流伯努利方程的应用

(1) 流量的测量——文丘里流量计　当管路中液体流经节流装置时，液流断面收缩，在收缩断面处流速增加，压力降低，使节流装置前后产生压差。现在我们以文丘里流量计为例，来推导计量流量的公式。文丘里流量计是用来计量有压管道流量的仪器，通常由渐缩段、喉道及渐扩段组成，如图 4-17 所示。取断面 1—1 和 2—2，计算点均取在管道上，基准面 0—0 置于管道下方某一固定位置，并取 $\alpha_1 = \alpha_2 = 1.0$，忽略摩擦，对 1—1、2—2 两个过流断面列总流伯努利方程，有

$$z_1+\frac{p_1}{\gamma}+\frac{v_1^2}{2g}=z_2+\frac{p_2}{\gamma}+\frac{v_2^2}{2g}$$

由连续性方程可得
$$v_1 A_1 = v_2 A_2$$

联立上面二式可得

$$v_1=\frac{1}{\sqrt{\left(\frac{A_1}{A_2}\right)^2-1}}\sqrt{2g\left[\left(z_1+\frac{p_1}{\rho g}\right)-\left(z_2+\frac{p_2}{\rho g}\right)\right]}=\frac{1}{\sqrt{\left(\frac{A_1}{A_2}\right)^2-1}}\sqrt{2g\Delta h}$$

故通过流量计的体积流量为

$$Q'=v_1 A_1=\frac{A_1}{\sqrt{\left(\frac{A_1}{A_2}\right)^2-1}}\sqrt{2g\Delta h}=\frac{\pi d_1^2 d_2^2}{4\sqrt{d_1^4-d_2^4}}\sqrt{2g\Delta h}$$

考虑到流体黏性的影响，上式右端需乘以一个流量修正系数，则

$$Q=\mu\frac{\pi d_1^2 d_2^2}{4\sqrt{d_1^4-d_2^4}}\sqrt{2g\Delta h} \tag{4-31}$$

(2) System with energy input or energy output

① Energy input—water pump

Choose section 1-1 and section 2-2 in Figure 4-18 (a), we can establish energy equation

$$z_1+\frac{p_1}{\gamma}+\frac{\alpha_1 v_1^2}{2g}+H_m=z_2+\frac{p_2}{\gamma}+\frac{\alpha_2 v_2^2}{2g}+h_{w_{1\text{-}2}}$$

where H_m—the increment energy of water per unit weight which is also called pump lift.

$$\frac{p_1}{\gamma}=0\text{m},\ \frac{p_2}{\gamma}=0\text{m},\ \frac{\alpha_1 v_1^2}{2g}=\frac{\alpha_2 v_2^2}{2g}=0\text{m}$$

$$H_m=(z_2-z_1)+h_{w_{1\text{-}2}}$$

Pump lift is
$$H_m=z+h_{w_{1\text{-}2}}$$

and valid power is
$$P_e=\gamma Q H_m$$

Let axes power be N, so efficiency is $\eta=\dfrac{\gamma Q H_m}{N}$

② Energy output—water turbine

$$z_1+\frac{p_1}{\gamma}+\frac{\alpha_1 v_1^2}{2g}-H_m=z_2+\frac{p_2}{\gamma}+\frac{\alpha_2 v_2^2}{2g}+h_{w_{1\text{-}2}}$$

where H_m—the increment energy of turbine per unit weight.

$$H_m=z-h_{w_{1\text{-}2}}$$

(2) 能量的输入或输出系统

① 能量的输入——水泵。取图 4-18 (a) 渐变流断面 1—1 和 2—2 建立能量方程

$$z_1+\frac{p_1}{\gamma}+\frac{\alpha_1 v_1^2}{2g}+H_m=z_2+\frac{p_2}{\gamma}+\frac{\alpha_2 v_2^2}{2g}+h_{w_{1\text{-}2}}$$

式中，H_m 为单位重量的水增加的能量，就是水泵的扬程。

$$\frac{p_1}{\gamma}=0\text{m},\ \frac{p_2}{\gamma}=0\text{m},\ \frac{\alpha_1 v_1^2}{2g}=\frac{\alpha_2 v_2^2}{2g}=0\text{m}$$

$$H_m=(z_2-z_1)+h_{w_{1\text{-}2}}$$

水泵的扬程为
$$H_m = z + h_{w_{1-2}}$$
有效功率为
$$P_e = \gamma Q H_m$$
令轴功率为 N，则效率为
$$\eta = \frac{\gamma Q H_m}{N}$$

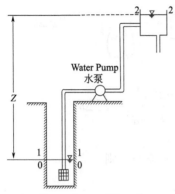
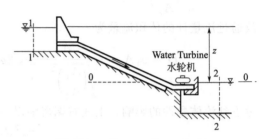

(a) Energy Input—Water Pump
(a) 能量的输入——水泵

(b) Energy Output—Water Turbine
(b) 能量的输出——水轮机

Figure 4-18 System with Energy Input or Energy Output
图 4-18 能量的输入与输出系统

② 能量的输出——水轮机

$$z_1 + \frac{p_1}{\gamma} + \frac{\alpha_1 v_1^2}{2g} - H_m = z_2 + \frac{p_2}{\gamma} + \frac{\alpha_2 v_2^2}{2g} + h_{w_{1-2}}$$

式中，H_m 为单位重量的水流给予水轮机的能量。

$$H_m = z - h_{w_{1-2}}$$

(3) Energy equation for conflux and distributary Assume there is a steady conflux as shown in Figure 4-19 (a), there is a conflux's surface ab. Applying continuity equation, we can obtain

$$Q_1 + Q_2 = Q_3$$

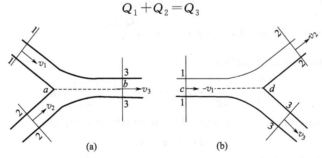

Figure 4-19 Energy Equation for Conflux (a) and Distributary (b)
图 4-19 汇流 (a) 与分流 (b) 的能量方程

According to the law of conservation and translation of energy, we can write energy equation for every section flow.

$$\left. \begin{array}{l} z_1 + \dfrac{p_1}{\gamma} + \dfrac{\alpha_1 v_1^2}{2g} = z_3 + \dfrac{p_3}{\gamma} + \dfrac{\alpha_3 v_3^2}{2g} + h_{w_{1-2}} \\[2mm] z_2 + \dfrac{p_2}{\gamma} + \dfrac{\alpha_2 v_2^2}{2g} = z_3 + \dfrac{p_3}{\gamma} + \dfrac{\alpha_3 v_3^2}{2g} + h_{w_{2-3}} \end{array} \right\} \quad (4\text{-}32)$$

Similarly, assume a steady distributary as shown in Figure 4-19 (b), there is a distributary's surface cd. Applying continuity equation yields

$$Q_1 = Q_2 + Q_3$$

According to the law of conservation and translation of energy, we can write energy equation for every section flow.

$$\left. \begin{array}{l} z_1 + \dfrac{p_1}{\gamma} + \dfrac{\alpha_1 v_1^2}{2g} = z_2 + \dfrac{p_2}{\gamma} + \dfrac{\alpha_2 v_2^2}{2g} + h_{w_{1\text{-}2}} \\ z_2 + \dfrac{p_2}{\gamma} + \dfrac{\alpha_2 v_2^2}{2g} = z_3 + \dfrac{p_3}{\gamma} + \dfrac{\alpha_3 v_3^2}{2g} + h_{w_{2\text{-}3}} \end{array} \right\} \quad (4\text{-}33)$$

(3) 有流量汇流或分流的能量方程 设有一恒定汇流，如图 4-19 (a) 所示。设想在汇流处做出汇流面 ab，每股总流的流量是不变的，且满足连续方程

$$Q_1 + Q_2 = Q_3$$

这样，根据能量守恒和转化定律就可分别写出每股总流的伯努利能量方程：

$$\left. \begin{array}{l} z_1 + \dfrac{p_1}{\gamma} + \dfrac{\alpha_1 v_1^2}{2g} = z_3 + \dfrac{p_3}{\gamma} + \dfrac{\alpha_3 v_3^2}{2g} + h_{w_{1\text{-}2}} \\ z_2 + \dfrac{p_2}{\gamma} + \dfrac{\alpha_2 v_2^2}{2g} = z_3 + \dfrac{p_3}{\gamma} + \dfrac{\alpha_3 v_3^2}{2g} + h_{w_{2\text{-}3}} \end{array} \right\} \quad (4\text{-}32)$$

同理，设有一恒定分流，如图 4-19 (b) 所示。设想在分流处做出汇流面 cd，每股总流的流量是不变的，且满足连续方程

$$Q_1 = Q_2 + Q_3$$

这样，根据能量守恒和转化定律就可分别写出每股总流的伯努利能量方程：

$$\left. \begin{array}{l} z_1 + \dfrac{p_1}{\gamma} + \dfrac{\alpha_1 v_1^2}{2g} = z_2 + \dfrac{p_2}{\gamma} + \dfrac{\alpha_2 v_2^2}{2g} + h_{w_{1\text{-}2}} \\ z_2 + \dfrac{p_2}{\gamma} + \dfrac{\alpha_2 v_2^2}{2g} = z_3 + \dfrac{p_3}{\gamma} + \dfrac{\alpha_3 v_3^2}{2g} + h_{w_{2\text{-}3}} \end{array} \right\} \quad (4\text{-}33)$$

(4) Energy change

【Sample Problem 4-1】 As shown in Figure 4-20, the stream discharges into the air at D. Let $a = b = c = 1\text{m}$, and the diameter of the vertical pipe is 100mm. If the loss of energy can be ignored, what are pressure heads at B, C and D?

Solution: Assume the datum line is across D and energy equation could be written between section A and D.

$$z_A + \frac{p_A}{\gamma} + \frac{v_A^2}{2g} = z_D + \frac{p_D}{\gamma} + \frac{v_D^2}{2g}$$

Because the area of water tank is far larger than that of water pipe, so we can consider that

$$\frac{v_A^2}{2g} = 0\text{m}$$

In addition, $z_A = 3\text{m}$, $\dfrac{p_A}{\gamma} = 0\text{m}$, $z_D = 0\text{m}$, $\dfrac{p_D}{\gamma} = 0\text{m}$.

So

$$3 + 0 + 0 = 0 + 0 + \frac{v_D^2}{2g}$$

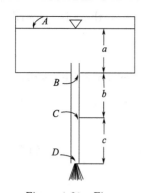

Figure 4-20 Figure for Sample Problem 4-1

图 4-20 例题 4-1 图

$$\frac{v_D^2}{2g}=3\text{m}$$

According to the continuity equation, $v_B A_B = v_C A_C = v_D A_D$
For $A_B = A_C = A_D$, then $v_B = v_C = v_D$
Energy equation could be written between section A and B.

$$z_A + \frac{p_A}{\gamma} + \frac{v_A^2}{2g} = z_B + \frac{p_B}{\gamma} + \frac{v_B^2}{2g}$$

$$3+0+0=2+\frac{p_B}{\gamma}+3, \quad \frac{p_B}{\gamma}=-2\text{m}$$

Energy equation could be written between section A and C.

$$z_A + \frac{p_A}{\gamma} + \frac{v_A^2}{2g} = z_C + \frac{p_C}{\gamma} + \frac{v_C^2}{2g}$$

$$3+0+0=1+\frac{p_C}{\gamma}+3, \quad \frac{p_C}{\gamma}=-1\text{m}$$

（4）能量的转化

【例题 4-1】 如图 4-20 所示，令 $a=b=c=1\text{m}$，水流在 D 点排入空气。垂直管的直径为 100mm，如忽略所有的能量损失，假设无水头损失，求 B、C、D 点各处的压强水头。

解：过 D 点做基准线，以过 A 及 D 的断面建立能量方程：

$$z_A + \frac{p_A}{\gamma} + \frac{v_A^2}{2g} = z_D + \frac{p_D}{\gamma} + \frac{v_D^2}{2g}$$

因为水箱断面比水管断面大得多，可以近似认为 $\frac{v_A^2}{2g}=0\text{m}$，又 $z_A=3\text{m}$，$\frac{p_A}{\gamma}=0\text{m}$，$z_D=0\text{m}$，$\frac{p_D}{\gamma}=0\text{m}$，则

$$3+0+0=0+0+\frac{v_D^2}{2g}$$

$$\frac{v_D^2}{2g}=3\text{m}$$

由连续性方程，$v_B A_B = v_C A_C = v_D A_D$，又 $A_B = A_C = A_D$，则 $v_B = v_C = v_D$。

以过 A 及 B 的断面建立能量方程：

$$z_A + \frac{p_A}{\gamma} + \frac{v_A^2}{2g} = z_B + \frac{p_B}{\gamma} + \frac{v_B^2}{2g}$$

$$3+0+0=2+\frac{p_B}{\gamma}+3, \quad \frac{p_B}{\gamma}=-2\text{m}$$

以过 A 及 C 的断面建立能量方程：

$$z_A + \frac{p_A}{\gamma} + \frac{v_A^2}{2g} = z_C + \frac{p_C}{\gamma} + \frac{v_C^2}{2g}$$

$$3+0+0=1+\frac{p_C}{\gamma}+3, \quad \frac{p_C}{\gamma}=-1\text{m}$$

(5) The height of pump installation

【Sample Problem 4-2】 As shown in Figure 4-21, a centrifugal pump draws 220L/s of water from the reservoir. Given that the vacuum height of inlet is 4.5m and the pipe diameter is 300mm. If the total friction head loss (h_w) is 1m, please calculate the most height (H_s) from the inlet axes of pump to water surface.

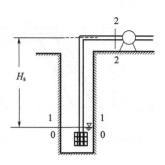

Figure 4-21 The Height of Pump Installation

图 4-21 水泵安装高度

Solution: Assume water surface 1-1 and pump inlet 2-2 are gradually-varying flow sections, and water surface is the datum line from 0-0. Energy equation could be written between section 1-1 and section 2-2.

$$z_1 + \frac{p_1}{\gamma} + \frac{v_1^2}{2g} = z_2 + \frac{p_2}{\gamma} + \frac{v_2^2}{2g} + h_{w_{1-2}}$$

Because the velocity of water surface is very little, we can consider that $\frac{v_1^2}{2g} \approx 0$.

And $z_1 = 0$, $z_2 = H_s$, $\frac{p_1}{\gamma} = 0\text{m}$, $\frac{p_2}{\gamma} = -4.5\text{m}$

$$v_2 = \frac{Q}{A_2} = \frac{0.22}{0.785 \times 0.3^2} = 3.11\text{m/s}$$

Substituting the data into energy equation

$$0 + 0 + 0 = H_s - 4.5 + 0.5 + 1.0$$

The height of pump installation $H_s = 3\text{m}$

(5) 水泵安装高度

【例题 4-2】 一台离心泵，抽水量为 220L/s，水泵进口允许真空度为 4.5m 水柱，水泵进口管径 $d=300$mm，如图 4-21 所示。从吸水滤头到水泵进口的水头损失 $h_w=1$m，试求能保证水泵吸水的进口轴线至水源水面的最大高度（称水泵的最大安装高度）H_s。

解：取水源水面（不是水管断面）、水泵进口为渐变流断面 1—1 和 2—2，以水池水面 0—0 为基准面，写出总流能量方程：

$$z_1 + \frac{p_1}{\gamma} + \frac{v_1^2}{2g} = z_2 + \frac{p_2}{\gamma} + \frac{v_2^2}{2g} + h_{w_{1-2}}$$

因水源水面流速较小，可取 $\frac{v_1^2}{2g} \approx 0$。又 $z_1 = 0$，$z_2 = H_s$，$\frac{p_1}{\gamma} = 0\text{m}$，$\frac{p_2}{\gamma} = -4.5\text{m}$，

$$v_2 = \frac{Q}{A_2} = \frac{0.22}{0.785 \times 0.3^2} = 3.11\text{m/s}$$

将已知数据代入总能量方程得

$$0 + 0 + 0 = H_s - 4.5 + 0.5 + 1.0$$

得水泵安装高度 $H_s = 3\text{m}$

(6) Siphon

【Sample Problem 4-3】 A siphon in Figure 4-22 draws water from the reservoir for irrigation, where $H=2$m and $d=100$mm. Assume head loss between the intake A and B is $10\frac{v^2}{2g}$ and the head loss between B and C is $2\frac{v^2}{2g}$. If the vacuum height does not exceed 7m, try to calculate: ①what is the most flux in the siphon? ②what is the height h between water surface and siphon exit?

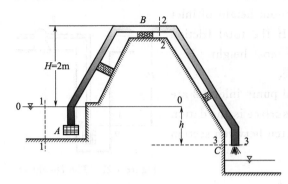

Figure 4-22 A Siphon for Irrigation
图 4-22 用于灌溉的虹吸管

Solution: ① Assume section 1-1, section 2-2 and section 3-3 are gradually-varying flow sections, and the water surface 0-0 is datum plane. Let $\alpha_1 = \alpha_2 = 1.0$, we can write energy equation between section 1-1 and section 2-2

$$z_1 + \frac{p_1}{\gamma} + \frac{v_1^2}{2g} = z_2 + \frac{p_2}{\gamma} + \frac{v_2^2}{2g} + h_{w_{1-2}}$$

Because the velocity of water surface is very little, we can consider that $\frac{v_1^2}{2g} \approx 0$.

$$z_1 = 0, \quad z_2 = 2\text{m}, \quad p_1 = 0, \quad \frac{p_2}{\gamma} = -7\text{m}$$

Substituting the data into energy equation

$$0 + 0 + 0 = 2 - 7 + \frac{v_2^2}{2g} + 10\frac{v^2}{2g}$$

Because $v = v_2$, then

$$v = \sqrt{\frac{5 \times 19.6}{11}} = 2.98 \text{m/s}$$

The most flux $\quad Q = \frac{\pi}{4}d^2 v = 0.785 \times 0.1^2 \times 2.98 = 23.4 \text{L/s}$

② Assume that section 3-3 is datum plane, and then we can write energy equation between section 2-2 and section 3-3.

$$z_2 = 2 + h, \quad z_3 = 0, \quad \frac{p_2}{\gamma} = -7\text{m}, \quad p_3 = 0$$

Let $\alpha_3 = 1.0$, substituting the data into energy equation

$$2 + h - 7 + \frac{v_2^2}{2g} = 0 + 0 + \frac{v_3^2}{2g} + 2\frac{v^2}{2g}$$

In this formula, $v_2 = v_3 = v$, so

$$h = 5 + 2 \times \frac{2.98^2}{19.6} = 5.91 \text{m}$$

The height h between water surface and siphon exit is 5.91m.

(6) 虹吸

【例题 4-3】有一从水池引水灌溉的虹吸管（图 4-22），管径 $d = 100$mm，管中心线最高处点 2 高出水池水面 2m，管段 AB 的点 1 至 2 的水头损失为 $10\frac{v^2}{2g}$，管段 BC 的点 2 至点 3 的水头损失为 $2\frac{v^2}{2g}$，若点 2 的真空度不超过 7m 水柱，试求：①虹吸管的最大流量为多少？②水池水面至虹吸管出口的高差 h 为多少？

解：①假设通过水池水面 1 点的过水断面 1—1、虹吸管顶部管末端断面 2—2 及虹吸管的出口断面 3—3 是渐变流断面。以水池水面 0—0 为基准面，取 $\alpha_1 = \alpha_2 = 1.0$，写出断面

1—1 和断面 2—2 的总能量方程：

$$z_1 + \frac{p_1}{\gamma} + \frac{v_1^2}{2g} = z_2 + \frac{p_2}{\gamma} + \frac{v_2^2}{2g} + h_{w_{1\text{-}2}}$$

水池水面流速很小，所以可认为 $\frac{v_1^2}{2g} \approx 0$。

又有 $\quad z_1 = 0$，$z_2 = 2\text{m}$，$p_1 = 0$，$\frac{p_2}{\gamma} = -7\text{m}$

将已知数据代入上式

$$0 + 0 + 0 = 2 - 7 + \frac{v_2^2}{2g} + 10\frac{v^2}{2g}$$

由连续性方程，管中 $v = v_2$，所以

$$v = \sqrt{\frac{5 \times 19.6}{11}} = 2.98 \text{m/s}$$

虹吸管最大流量

$$Q = \frac{\pi}{4}d^2 v = 0.785 \times 0.1^2 \times 2.98 = 23.4 \text{L/s}$$

② 以管的出口断面 3—3 为基准面，写出断面 2—2 和断面 3—3 的能量方程：$z_2 = 2 + h$，$z_3 = 0$，$\frac{p_2}{\gamma} = -7\text{m}$，$p_3 = 0$。取 $\alpha_3 = 1.0$，将已知数据代入能量方程得

$$(2+h) - 7 + \frac{v_2^2}{2g} = 0 + 0 + \frac{v_3^2}{2g} + 2\frac{v^2}{2g}$$

式中 $v_2 = v_3 = v$，则有

$$h = 5 + 2 \times \frac{2.98^2}{19.6} = 5.91\text{m}$$

水池水面至虹吸管出口的高差 h 为 5.91m。

4.8 Momentum Equation and Its Application

4.8.1 Momentum Equation

The momentum principle can be derived from Newton's second law

$$\sum \vec{F} = \frac{\mathrm{d}(m\vec{v})}{\mathrm{d}t}$$

This states that the sum of the external forces $\vec{F}$ on a body of fluid or system is equal to the rate of change of linear momentum $m\vec{v}$ of that body or system. $\vec{F}$ and $\vec{v}$ represent vectors, and the change in momentum must be in the same direction as force. We can also express the principle as

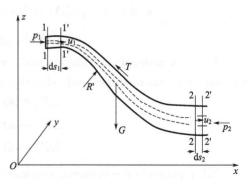

Figure 4-23 The Derivation of Momentum Equation

图 4-23 动量方程的推导

$$\Sigma \vec{F} \times \Delta t = \Delta(m\vec{v})$$

Impulse equals change of momentum. That is called impulse momentum principle.

$$\text{Impulse} = \text{Change of Momentum}$$

As shown in Figure 4-23, let p_1, u_1, dA_1 and p_2, u_2, dA_2 be pressure, velocity and the area at 1 and 2 sections, respectively. The fixed control volume lies between section 1 and 2. The moving fluid system consists of the fluid mass contained at time t in the control volume. During a short time interval dt, assume that the fluid contained in the control volume moves a short distance ds_1 at section 1 and ds_2 at section 2.

The increment of momentum between 1-1' and 2-2'

$$\Delta(m\vec{v}) = \rho ds_2 dA_2 \vec{u}_2 - \rho ds_2 dA_2 \vec{u}_1 = \rho dQ dt (\vec{u}_2 - \vec{u}_1)$$

Applying the impulse momentum principle

$$\Sigma \vec{F} \times dt = \rho dQ dt (\vec{u}_2 - \vec{u}_1)$$

Resultant force for element flow

$$\Sigma \vec{F} = \rho dQ (\vec{u}_2 - \vec{u}_1)$$

Integrating

$$\Sigma \vec{F} = \int_{A_2} \rho \vec{u}_2^2 dA_2 - \int_{A_1} \rho \vec{u}_1^2 dA_1$$

$$\beta = \frac{\int_A u^2 dA}{v^2 A}$$

$$\int_A u^2 dA = \int_A (v + \Delta u)^2 dA = \int_A (v^2 + 2v\Delta u + \Delta u^2) dA = v^2 A + \int_A 2v\Delta u dA + \int_A \Delta u^2 dA$$

For

$$\int_A (2v\Delta u) dA = 2v \int_A \Delta u dA = 0, \quad \int_A \Delta u^2 dA \geqslant 0$$

$$\int_A u^2 dA = v^2 A + \int_A \Delta u^2 dA$$

So

$$\beta = \frac{\int_A u^2 dA}{v^2 A} \geqslant 1$$

$$\int_{A_1} \rho u_1^2 dA_1 = \beta_1 \rho v_1^2 A_1 = \beta_1 \rho v_1 Q, \quad \int_{A_2} \rho u_2^2 dA_2 = \beta_2 \rho v_2^2 A_2 = \beta_2 \rho v_2 Q$$

$$\Sigma \vec{F} = \rho Q (\beta_2 \vec{v}_2 - \beta_1 \vec{v}_1) \tag{4-34}$$

This is a momentum equation and a vector equation. We can also express it as scalar equations in terms of forces and velocities in the x, y and z directions, respectively.

$$\left. \begin{array}{l} \Sigma F_x = \rho Q (\beta_2 v_{2x} - \beta_1 v_{1x}) \\ \Sigma F_y = \rho Q (\beta_2 v_{2y} - \beta_1 v_{1y}) \\ \Sigma F_z = \rho Q (\beta_2 v_{2z} - \beta_1 v_{1z}) \end{array} \right\} \tag{4-35}$$

$\Sigma \vec{F}$ represents the vectorial summation of all forces acting on the fluid mass in the control volume, including gravity force G, shear force T, pressure force P exerted by fluids, and pressure force R exerted by the solid boundaries in contact with the fluid mass.

Resultant force is the sum of G, T, P and R

$$\sum \vec{F} = \vec{G} + \vec{T} + \vec{P} + \vec{R}$$

Assume $\vec{r}$ is the radius vector from a reference point to acting point of fluid velocity vector $\vec{u}$.

Then use this vector $\vec{r}$ to multiply both sides of momentum equation. The moment of momentum equation for element flow is

$$r\vec{F} = \rho dQ \ (r_2 \vec{u}_2 - r_1 \vec{u}_1)$$

Integrating

$$\sum r\vec{F} = \rho Q \ (\beta_2 r_2 \vec{v}_2 - \beta_1 r_1 \vec{v}_1) \tag{4-36}$$

This is a moment of momentum equation and a vectorial equation. It shows that the resultant momentum of resultant forces equals to the fluid moment of momentum difference between fluids flowing out of and flowing into the control volume.

We can also express it as scalar equations in terms of forces and velocities in the x, y and z directions, respectively.

$$\left. \begin{array}{l} \sum rF_x = \rho Q \ (\beta_2 r_2 v_{2x} - \beta_1 r_1 v_{1x}) \\ \sum rF_y = \rho Q \ (\beta_2 r_2 v_{2y} - \beta_1 r_1 v_{1y}) \\ \sum rF_z = \rho Q \ (\beta_2 r_2 v_{2z} - \beta_1 r_1 v_{1z}) \end{array} \right\} \tag{4-37}$$

4.8 动量方程及其应用

4.8.1 动量方程

动量原理可以从牛顿第二运动定律推导得出

$$\sum \vec{F} = \frac{d \ (m\vec{v})}{dt}$$

作用在流体的控制体或系统上的外力等于动量的变化率，$\vec{F}$ 和 $\vec{v}$ 表示矢量，动量的变化必须与力的方向一致。外力也可以表示为

$$\sum \vec{F} \times \Delta t = \Delta \ (m\vec{v})$$

冲量等于动量的变化，这也称为冲量定理

$$\text{冲量} = \text{动量的增量}$$

在图 4-23 中，p_1、u_1、dA_1 和 p_2、u_2、dA_2 分别表示 1、2 两个断面的压强、速度及面积。控制体位于 1、2 两个断面之间，控制体内的流体经过 dt 时段分别在 1、2 断面移动了 ds_1 和 ds_2。

在断面 1—1′和断面 2—2′之间的动量增量为

$$\Delta(m\vec{v}) = \rho ds_2 dA_2 \vec{u}_2 - \rho ds_2 dA_2 \vec{u}_1 = \rho dQ dt \ (\vec{u}_2 - \vec{u}_1)$$

应用冲量定理

$$\sum \vec{F} \times dt = \rho dQ dt \ (\vec{u}_2 - \vec{u}_1)$$

元流的合力

$$\sum \vec{F} = \rho dQ \ (\vec{u}_2 - \vec{u}_1)$$

积分得

$$\sum \vec{F} = \int_{A_2} \rho \vec{u}_2^2 dA_2 - \int_{A_1} \rho \vec{u}_1^2 dA_1$$

$$\beta = \frac{\int_A u^2 \mathrm{d}A}{v^2 A}$$

$$\int_A u^2 \mathrm{d}A = \int_A (v+\Delta u)^2 \mathrm{d}A = \int_A (v^2 + 2v\Delta u + \Delta u^2) \mathrm{d}A = v^2 A + \int_A 2v\Delta u \mathrm{d}A + \int_A \Delta u^2 \mathrm{d}A$$

因为
$$\int_A (2v\Delta u)\mathrm{d}A = 2v \int_A \Delta u \mathrm{d}A = 0, \quad \int_A \Delta u^2 \mathrm{d}A \geqslant 0$$

$$\int_A u^2 \mathrm{d}A = v^2 A + \int_A \Delta u^2 \mathrm{d}A$$

所以
$$\beta = \frac{\int_A u^2 \mathrm{d}A}{v^2 A} \geqslant 1$$

$$\int_{A_1} \rho u_1^2 \mathrm{d}A_1 = \beta_1 \rho v_1^2 A_1 = \beta_1 \rho v_1 Q, \quad \int_{A_2} \rho u_2^2 \mathrm{d}A_2 = \beta_2 \rho v_2^2 A_2 = \beta_2 \rho v_2 Q$$

$$\sum \vec{F} = \rho Q (\beta_2 \vec{v}_2 - \beta_1 \vec{v}_1) \tag{4-34}$$

这就是动量方程, 是一个矢量方程, 也可以在 x、y、z 三个方向分别表示成力和速度的标量:

$$\left.\begin{array}{l}\sum F_x = \rho Q (\beta_2 v_{2x} - \beta_1 v_{1x}) \\ \sum F_y = \rho Q (\beta_2 v_{2y} - \beta_1 v_{1y}) \\ \sum F_z = \rho Q (\beta_2 v_{2z} - \beta_1 v_{1z})\end{array}\right\} \tag{4-35}$$

$\sum \vec{F}$ 表示作用在控制体上的所有力的矢量之和, 包括重力 G、剪切力 T、流体的压力 P 及固体边界对流体的作用力 R。

合力为 G、T、P、R 之和

$$\sum \vec{F} = \vec{G} + \vec{T} + \vec{P} + \vec{R}$$

设 $\vec{r}$ 为某参考点至流体速度矢量 $\vec{u}$ 的作用点的矢径, 则用此矢量 $\vec{r}$ 对动量方程两端进行矢性积运算, 可得元流的动量矩方程为

$$r\vec{F} = \rho \mathrm{d}Q (r_2 \vec{u}_2 - r_1 \vec{u}_1)$$

积分得

$$\sum r\vec{F} = \rho Q (\beta_2 r_2 \vec{v}_2 - \beta_1 r_1 \vec{v}_1) \tag{4-36}$$

这就是总流的动量矩方程, 为矢量方程。它表明, 单位时间内从控制面流出的动量矩减去流入控制面的动量矩, 等于作用在控制体上所有的外力矩之和。

也可以在 x、y、z 三个方向表示成力和速度的标量, 即

$$\left.\begin{array}{l}\sum rF_x = \rho Q (\beta_2 r_2 v_{2x} - \beta_1 r_1 v_{1x}) \\ \sum rF_y = \rho Q (\beta_2 r_2 v_{2y} - \beta_1 r_1 v_{1y}) \\ \sum rF_z = \rho Q (\beta_2 r_2 v_{2z} - \beta_1 r_1 v_{1z})\end{array}\right\} \tag{4-37}$$

4.8.2 Application of Momentum Equation

【**Sample Problem 4-4**】 As shown in Figure 4-24, a horizontal pipe transports water from section 1 to 2, where $d_1 = 1.5\mathrm{m}$, $d_2 = 1.0\mathrm{m}$, $p_1 = 39.2 \times 10^4 \mathrm{Pa}$, and $Q = 1.8 \mathrm{m}^3/\mathrm{s}$. If losses are neglected, compute the force R acting on the pipe wall.

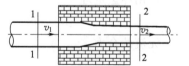

Figure 4-24 Figure for Sample Problem 4-4

图 4-24 例题 4-4 图

Solution: Assume that R' is the force pipe wall acting on the water. The direction of R' is horizontal and leftward. The momentum equation could be written as follows.

$$p_1 A_1 - p_2 A_2 - R' = \rho Q (v_2 - v_1)$$

And
$$v_1 = \frac{4Q}{\pi d_1^2} = \frac{4 \times 1.8}{\pi \times 1.5^2} = 1.02 \text{m/s}$$

$$v_2 = \frac{4Q}{\pi d_2^2} = \frac{4 \times 1.8}{\pi \times 1.0^2} = 2.29 \text{m/s}$$

According to energy equation, we can obtain

$$p_2 = \gamma \left(\frac{p_1}{\gamma} + \frac{v_1^2}{2g} - \frac{v_2^2}{2g} \right) = 9800 \times \left(\frac{39.2 \times 10^4}{9800} + \frac{1.02^2}{2 \times 9.8} - \frac{2.29^2}{2 \times 9.8} \right) = 39.0 \times 10^4 \text{Pa}$$

Then $R' = p_1 A_1 - p_2 A_2 - \rho Q (v_2 - v_1)$

$$= 39.2 \times 10^4 \times \frac{\pi \times 1.5^2}{4} - 39.0 \times 10^4 \times \frac{\pi \times 1.0^2}{4} - 1000 \times 1.8 \times (2.29 - 1.02)$$

$$= 384.1 \times 10^3 \text{N} = 384.1 \text{kN}$$

So $R = R' = 384.1$ kN. The direction of R is horizontal and rightward.

4.8.2 动量方程的应用

【例题 4-4】 有一段水平输水管，如图 4-24 所示。已知 $d_1 = 1.5$m，$d_2 = 1.0$m，$p_1 = 39.2 \times 10^4$Pa，$Q = 1.8$m^3/s，水流由过流断面 1—1 流到过流断面 2—2，若不计能量损失，试求作用在该段管壁上的轴向力 R。

解：假设管壁作用于水体的力为 R'，方向为水平向左，由动量方程可得：

$$p_1 A_1 - p_2 A_2 - R' = \rho Q (v_2 - v_1)$$

而
$$v_1 = \frac{4Q}{\pi d_1^2} = \frac{4 \times 1.8}{\pi \times 1.5^2} = 1.02 \text{m/s}$$

$$v_2 = \frac{4Q}{\pi d_2^2} = \frac{4 \times 1.8}{\pi \times 1.0^2} = 2.29 \text{m/s}$$

由能量方程可得：

$$p_2 = \gamma \left(\frac{p_1}{\gamma} + \frac{v_1^2}{2g} - \frac{v_2^2}{2g} \right) = 9800 \times \left(\frac{39.2 \times 10^4}{9800} + \frac{1.02^2}{2 \times 9.8} - \frac{2.29^2}{2 \times 9.8} \right) = 39.0 \times 10^4 \text{Pa}$$

则 $R' = p_1 A_1 - p_2 A_2 - \rho Q (v_2 - v_1)$

$$= 39.2 \times 10^4 \times \frac{\pi \times 1.5^2}{4} - 39.0 \times 10^4 \times \frac{\pi \times 1.0^2}{4} - 1000 \times 1.8 \times (2.29 - 1.02)$$

$$= 384.1 \times 10^3 \text{N} = 384.1 \text{kN}$$

因此 $R = R' = 384.1$ kN，方向与 R' 相反，即 R 的方向为水平向右。

【Sample Problem 4-5】 The axes of a changing size syphon (Figure 4-25) are on the same horizontal plane. The corner is $\alpha = 60°$ and diameter turns from $d_A = 200$mm into $d_B = 150$mm. When the flux is $Q = 0.1$m^3/s, and the pressure is $p_A = 18$kN/m^2, try to calculate the acting force of the stream to segment AB. The loss of syphon segment can be neglected.

Solution: To calculate the acting force of fluids and borders, normally the continuous equation, energy equation and momentum equation should be used united.

① Take out segment AB as a partition and prescribe the positive direction of coordinate. Assume the direction of reactive force R_x and R_y, and then momentum equations in x

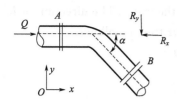

Figure 4-25 Figure for Sample Problem 4-5

图 4-25 例题 4-5 图

and y directions can be written.

$$\sum F_x = \rho Q \ (v_{Bx} - v_{Ax})$$
$$\sum F_y = \rho Q \ (v_{By} - v_{Ay})$$

Substituting outside forces and velocities into the equations above yields

$$p_A \frac{\pi}{4} d_A^2 - p_B \frac{\pi}{4} d_B^2 \cos\alpha + R_x = \rho Q \ (v_B \cos\alpha - v_A) \quad (1)$$

$$p_B \frac{\pi}{4} d_B^2 \sin\alpha + R_y = \rho Q \ (-v_B \sin\alpha - 0) \quad (2)$$

② Use energy equation to calculate p_B

$$p_B = p_A + \gamma \left(\frac{v_A^2}{2g} - \frac{v_B^2}{2g} \right) = 18000 + 9800 \times \left(\frac{3.18^2}{2 \times 9.8} - \frac{5.66^2}{2 \times 9.8} \right) = 7.03 \text{kN/m}^2$$

③ Use continuous equation to calculate v_A and v_B

$$v_A = \frac{4Q}{\pi d_A^2} = \frac{4 \times 0.1}{\pi \times 0.2^2} = 3.18 \text{m/s}$$

$$v_B = \frac{4Q}{\pi d_B^2} = \frac{4 \times 0.1}{\pi \times 0.15^2} = 5.66 \text{m/s}$$

Substituting the data above into Equation (1) and Equation (2), then obtain

$$R_x = -0.538 \text{kN}, \quad R_y = -0.598 \text{kN}$$

$$R = \sqrt{R_x^2 + R_y^2} = 0.804 \text{kN}, \quad \theta = \tan^{-1} \frac{R_y}{R_x} = 48.02°$$

The acting force that fluid acts on siphon equals to the reverse force of siphon, but the direction is reverse, namely $R' = -R$.

【例题 4-5】 一变径弯管（图 4-25），轴线位于同一水平面，转角 $\alpha = 60°$，直径由 $d_A = 200$mm 变为 $d_B = 150$mm，在流量 $Q = 0.1\text{m}^3/\text{s}$ 时，压强 $p_A = 18\text{kN/m}^2$，求水流对 AB 段弯管的作用力。不计弯管段的水头损失。

解：求解流体与边界的作用力问题，一般需要联合使用连续性方程、能量方程和动量方程。

① 将流段 AB 作为隔离体取出，规定坐标正方向，假定弯管反作用力 R_x 和 R_y 的方向，写出 x 和 y 两个方向的动量方程：

$$\sum F_x = \rho Q \ (v_{Bx} - v_{Ax})$$
$$\sum F_y = \rho Q \ (v_{By} - v_{Ay})$$

把外力和流速代入其中得

$$p_A \frac{\pi}{4} d_A^2 - p_B \frac{\pi}{4} d_B^2 \cos\alpha + R_x = \rho Q \ (v_B \cos\alpha - v_A) \quad (1)$$

$$p_B \frac{\pi}{4} d_B^2 \sin\alpha + R_y = \rho Q \ (-v_B \sin\alpha - 0) \quad (2)$$

② 由能量方程可得到 p_B。

$$p_B = p_A + \gamma \left(\frac{v_A^2}{2g} - \frac{v_B^2}{2g} \right) = 18000 + 9800 \times \left(\frac{3.18^2}{2 \times 9.8} - \frac{5.66^2}{2 \times 9.8} \right) = 7.03 \text{kN/m}^2$$

③ 由连续性方程可得到 v_A 和 v_B。

$$v_A = \frac{4Q}{\pi d_A^2} = \frac{4 \times 0.1}{\pi \times 0.2^2} = 3.18 \text{m/s}$$

$$v_B = \frac{4Q}{\pi d_B^2} = \frac{4 \times 0.1}{\pi \times 0.15^2} = 5.66 \text{m/s}$$

将以上数据代入式(1) 和式(2)，可求得：

$$R_x = -0.538 \text{kN}, \quad R_y = -0.598 \text{kN}$$

$$R = \sqrt{R_x^2 + R_y^2} = 0.804 \text{kN}, \quad \theta = \tan^{-1} \frac{R_y}{R_x} = 48.02°$$

流体对弯管的作用力和弯管的反作用力大小相等，方向相反，即 $R' = -R$。

Exercises 习题

4-1 A Pitot tube with a coefficient of 0.98 is used to measure the velocity of water at the center of a pipe at 5.00m/s. What is the magnitude, in kPa, of the difference between the total pressure and the static pressure?

4-1 利用修正系数为 0.98 的皮托管测得管道中心点的流速为 5.00m/s，那么全压与静压之间的差值为多少？以 kPa 表示。

4-2 The diameter of siphon is 150mm. The sketch is shown in Figure. The diameter of nozzle exit d_2 is 50mm. Compute the flux and the pressure at point A, B, C and D. Assume no head loss.

4-2 已知虹吸管的直径 $d_1 = 150$mm，布置形式如图所示，喷嘴出口直径 $d_2 = 50$mm，不计水头损失，求虹吸管的输水流量及管中 A、B、C、D 各点压强值。

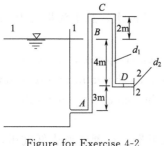

Figure for Exercise 4-2

习题 4-2 图

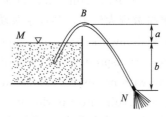

Figure for Exercise 4-3

习题 4-3 图

4-3 As shown in Figure, assume $a = 1$m, $b = 4$m, and the flow is frictionless in the siphon. Calculate the rate of discharge in m^3/s and the pressure head at B if the pipe has a uniform diameter of 150mm.

4-3 如图所示，假设 $a = 1$m，$b = 4$m，在管内流动无摩擦。求以 m^3/s 计的排放量和在 B 处的压强水头，假设管子的直径是均匀的，为 150mm。

4-4 Assume there is friction head loss in the siphon of Figure for Exercise 4-3, where $a = 1$m, $b = 4$m. The loss between the intake and B is 0.6m, between B and N is 0.9m. What is the rate of discharge and pressure head at B when the diameter is 150mm?

4-4 如习题 4-3 图所示，假设虹吸管内有摩擦水头损失，其中 $a = 1$m，$b = 4$m，在入口处与 B 点之间的损失是 0.6m，B 点和 N 点之间的损失是 0.9m。当直径是 150mm 时，排放量及 B 点的压强水头是多少？

4-5 In Figure, let $a=7.5$m, $b=c=15$m, and $d=300$mm. All the losses of energy are to be ignored when the stream discharging into the air at E with a diameter of 80mm. What are pressure heads at B, C and D if the diameter of the vertical pipe is 120mm?

4-5 如图所示，令 $a=7.5$m，$b=c=15$m，$d=300$mm，水流在 E 点排入空气中，E 点直径为 80mm，且忽略所有的能量损失。如果垂直管的直径是 120mm，B、C、D 点各处的压强水头是多少？

4-6 Find the flow rate of the two-dimensional channel shown in Figure. Assume no head loss.

4-6 如图所示，求二维河道的流量，假设不存在水头损失。

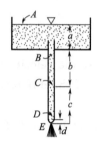

Figure for Exercise 4-5
习题 4-5 图

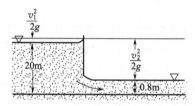

Figure for Exercise 4-6
习题 4-6 图

4-7 When the pump in Figure draws 220m³/h of water at 20℃ from the reservoir, the total friction head loss is 5m. The flow discharges through a nozzle to the air. Estimate the pump power in kW.

4-7 在图中，水泵以 220m³/h 的流量从水库中抽取 20℃ 的水，总水头损失为 5m，流体通过一个喷嘴排放到大气中，估计水泵功率（kW）。

4-8 A necked-down section in a pipe flow, called a water ejector, develops a low throat pressure that can aspirate fluid upward from a reservoir, as shown in Figure. Using Bernoulli's equation with no losses, derive an expression for the velocity v_1 just sufficient to bring fluid in the reservoir into the throat.

4-8 如图所示，一个收缩断面呈颈状的管流，称为水射器，在颈处形成低气压，可以从水池中向上吸液体。应用伯努利方程，假设没有损失，推导出一个足以使流体从水池进入颈处的速度 v_1 的表达式。

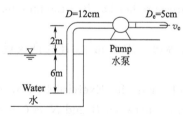

Figure for Exercise 4-7
习题 4-7 图

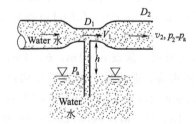

Figure for Exercise 4-8
习题 4-8 图

4-9 A fire hose, nozzle and pump are shown in Figure. The pressure at point A is 2atm (gage pressure). The diameter of the hose and the nozzle are 50mm and 20mm, respectively. The head loss of the fire hose and the nozzle are 0.5m and 0.1m, respectively. Try to calculate the velocity at exit, flow rate of pump and the pressure at point B.

4-9 一救火水龙带，喷嘴和泵的相对位置如图所示。泵出口压力（A 点压力）为 2atm（表压），泵排出管断面直径为 50mm，喷嘴出口 C 的直径为 20mm，水龙带的水头损失设为 0.5m，喷嘴水头损失为 0.1m。试求喷嘴出口流速、泵的排量及 B 点压力。

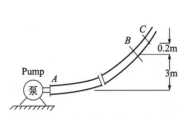

Figure for Exercise 4-9
习题 4-9 图

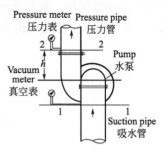

Figure for Exercise 4-10
习题 4-10 图

4-10 In Figure, the pump draws water from the reservoir, and $d_1 = 200$mm, $d_2 = 150$mm, $Q = 0.06$m^3/s, and $h = 0.5$m. The reading in vacuum meter of inlet is 4m H$_2$O. The reading in pressure meter of outlet is 2atm. (1) Please calculate the pump height H_m. (2) If the axes power is 25 horse powers, what is the pump efficiency η?

4-10 测定水泵扬程的装置如图所示。已知水泵吸水管直径 $d_1=200$mm，压水管直径 $d_2=150$mm，流量 $Q=0.06$m^3/s，水管与两表连接的测压孔位置之间的高差 $h=0.5$m，水泵进口真空表读数为 4m 水柱，水泵出口压力表读数为 2atm（工程大气压）。试求此时的水泵扬程 H_m。若同时测得水泵的轴功率 $N=25$hp（马力），试求水泵的效率 η。

4-11 Water at 20℃ exits to the standard sea-level atmosphere through the split nozzle in Figure. Duct areas are $A_1=0.02$m^2 and $A_2=A_3=0.008$m^2. If $p_1=135$kPa (absolute) and the flow rate is $Q_2=Q_3=275$m^3/h, compute the force on the flange bolts at section 1.

4-11 在图中，20℃的水通过分离喷嘴喷射到标准海平面大气中，管道面积 $A_1=0.02$m^2，$A_2=A_3=0.008$m^2。如果 $p_1=135$kPa（绝对压强），流量 $Q_2=Q_3=275$m^3/h，计算断面 1 处法兰螺栓上的作用力。

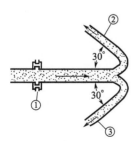

Figure for Exercise 4-11
习题 4-11 图

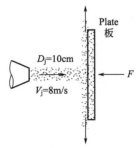

Figure for Exercise 4-12
习题 4-12 图

4-12 The water jet in Figure strikes normal to a fixed plate. Neglect gravity and friction, and compute the force F in Newtons required to hold the plate fixed.

4-12 如图所示，水射流到固定板上，忽略重力和摩擦力，计算使板固定所需的力 F 的大小（N）。

4-13 In the spillway flow shown in Figure, the flow is assumed uniform and hydrostatic at sections 1 and 2. If losses are neglected, compute (a) v_2 and (b) the force per unit width of the water on the spillway.

4-13 如图所示的溢流坝，假定水流在 1 和 2 断面处为均匀流及符合静水力学。如果忽略损失，计算 v_2 及作用在单位宽度溢流坝上的力。

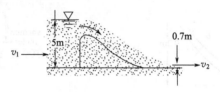

Figure for Exercise 4-13

习题 4-13 图

Chapter 5　Dimensional Analysis and Similarity Principle
第5章　量纲分析和相似原理

5.1　Dimensions and Units

5.1.1　Dimensions

The property and the type of physical quantity are expressed by dimensions. [Length] means the dimension of length and not a particular length with a definite numerical value. For conciseness, length is abbreviated to L and the dimension of length is written as [L]. Similarly [T] is used for the dimension of time, [M] for the dimension of mass. The ternary basic dimensions are [L], [T] and [M] in fluid mechanics.

The derivation dimension is composed of [L], [T] and [M]. For example, the dimension of velocity is $[LT^{-1}]$ and the dimension of force is $[LT^{-2}M]$. The dimensions of physical quantities used in fluid mechanics are shown in Table 5-1. Usually, the dimension will be expressed as follows.

$$\dim \chi = [L^\alpha T^\beta M^\gamma] \tag{5-1}$$

① when $\alpha \neq 0$, $\beta = 0$ and $\gamma = 0$, the physical quantity is a geometry quantity.
② when $\beta \neq 0$ and $\gamma = 0$, the physical quantity is a kinematics quantity.
③ when $\gamma \neq 0$, the physical quantity is a dynamics quantity.

5.1.2　Units

To complete the description of the physical quantity, it is necessary to know the magnitude of each dimension. The magnitude and size are expressed by units. A length would be measured in terms of a standardized unit of length, such as meter. Mass would be measured in terms of a standardized unit of mass, such as kilogram. The units of time are day, hour, minute and second, etc. Units of physical quantities used in fluid mechanics are shown in Table 5-1.

5.1.3　Dimensionless Quantities

When describing an object, we sometimes use quantities which are dimensionless. When $\alpha = 0$, $\beta = 0$ and $\gamma = 0$, Equation (5-1) will become

$$\dim \chi = [L^0 T^0 M^0] = 1 \tag{5-2}$$

So a dimensionless quantity is also called a quantity of dimension one. Such quantities are ratios comparing one quantity with another of the same kind and their numerical values are independent of the system of units employed. For example, $l = L/L_0$ and $\Theta = t/T$ are all dimensionless quantities. In addition, Reynolds number is also a dimensionless quantity.

$$Re = vd/\nu$$

where v — velocity and its dimension is $[LT^{-1}]$;

d — pipe diameter and its dimension is $[L]$;

ν — kinematic viscosity and its dimension is $[L^2 T^{-1}]$.

$$\dim Re = [LT^{-1}][L]/[L^2 T^{-1}] = 1$$

Table 5-1 Dimensions and Units of Physical Quantities Used in Fluid Mechanics

表 5-1 流体力学中的常用量纲和单位

	Physical quantities 物理量	Symbols 符号	Dimensions 量纲	Units 单位
Geometry quantities 几何学的量	Length 长度	L	L	m
	Area 面积	A	L^2	m^2
	Volume 体积	V	L^3	m^3
	Water head 水头	H	L	m
	Moment of area 面积矩	I	L^4	m^4
Kinematics quantities 运动学的量	Time 时间	t	T	s
	Velocity 速度	v	LT^{-1}	m/s
	Acceleration 加速度	a	LT^{-2}	m/s^2
	Angular rotation speed 角转速	ω	T^{-1}	rad/s
	Discharge 流量	Q	$L^3 T^{-1}$	m^3/s
	Kinematic viscosity 运动黏度	ν	$L^2 T^{-1}$	m^2/s
Dynamics quantities 动力学的量	Mass 质量	m	M	kg
	Density 密度	ρ	ML^{-3}	kg/m^3
	Force 力	F	MLT^{-2}	N
	Dynamic viscosity 动力黏度	μ	$ML^{-1} T^{-1}$	Pa·s
	Pressure 压强	p	$ML^{-1} T^{-2}$	Pa
	Shear stress 切应力	τ	$ML^{-1} T^{-2}$	Pa
	Elastic modulus 弹性模量	E	$ML^{-1} T^{-2}$	Pa
	Surface tension 表面张力	σ	MT^{-1}	N/m
	Momentum 动量	p	MLT^{-1}	kg·m/s
	Energy 能	W	$ML^2 T^{-2}$	J
	Power 功率	P	$ML^2 T^{-3}$	W
	Mass concentration 质量浓度	C	ML^{-3}	kg/m^3

5.1 量纲和单位

5.1.1 量纲

量纲用来表示物理量的性质和类别。[Length] 表示长度量纲，而不是有明确数量值的详细的长度。为了简洁，Length 可以简写为 L，长度量纲就可以写为 [L]。同样，[T] 表示时间量纲，[M] 表示质量量纲。流体力学中的三个基本量纲为 [L]、[T] 和 [M]。

由基本量纲组成的量纲称为导出量纲，如速度 v 的量纲为 $[LT^{-1}]$，力 F 的量纲为 $[LT^{-2} M]$。表 5-1 是流体力学中的常用量纲，通常量纲可以表示为

$$\dim \chi = [L^\alpha T^\beta M^\gamma] \tag{5-1}$$

① 当 $\alpha \neq 0$、$\beta = 0$、$\gamma = 0$ 时，为几何学的量。

② 当 $\beta \neq 0$、$\gamma = 0$ 时，为运动学的量。

③ 当 $\gamma \neq 0$ 时，为动力学的量。

5.1.2 单位

为了全面地描述物理量，很有必要知道每一量纲的数量，单位用来表示物理量的数量和大小。长度的标准单位是米，质量的标准单位是千克，时间可以用天、小时、分钟和秒来表示。流体力学中常用量的单位见表 5-1。

5.1.3 无量纲量

描述一个对象时，我们有时使用无量纲量。当 $\alpha=0$、$\beta=0$、$\gamma=0$ 时，代入式 (5-1) 可得：

$$\dim \chi = [L^0 T^0 M^0] = 1 \tag{5-2}$$

因此，无量纲量也称为量纲为一的量。这些物理量是一个量和另一个相同类型量之比，它们的数值不受单位的影响，如，$l=L/L_0$ 或 $\Theta=t/T$。除此之外，雷诺数也是一个无量纲量。

$$Re = vd/\nu$$

式中，v 为速度，其量纲是 $[LT^{-1}]$；d 为管径，其量纲是 $[L]$；ν 为运动黏度，其量纲是 $[L^2 T^{-1}]$。

$$\dim Re = [LT^{-1}][L]/[L^2 T^{-1}] = [1]$$

5.2 Dimensional Homogeneity

For an equation describing a physical phenomenon to be true, the two sides must be equal both numerically and dimensionally. An equation describing a physical phenomenon will be true only if all the terms are of the same kind and have the same dimensions. Thus the equation is considered to be dimensionally homogeneous. For example,

$$z_1 + \frac{p_1}{\gamma} + \frac{\alpha_1 v_1^2}{2g} = z_2 + \frac{p_2}{\gamma} + \frac{\alpha_2 v_2^2}{2g} + h_w$$

The dimension of every term in Bernoulli's equation is $[L]$.

5.2 量纲和谐原理

对于一个描述物理现象的正确方程，方程两边在数量和量纲上应分别相等。假如方程中的各项是相同类型，并且有相同的量纲，则这个方程能够正确地描述物理现象，就可以认为量纲和谐了。比如

$$z_1 + \frac{p_1}{\gamma} + \frac{\alpha_1 v_1^2}{2g} = z_2 + \frac{p_2}{\gamma} + \frac{\alpha_2 v_2^2}{2g} + h_w$$

伯努利方程中每一项的量纲均为 $[L]$。

5.3 Rayleigh's Method and Buckingham's π Theorem

5.3.1 Rayleigh's Method

If the physical quantity y depends upon the variables x_1, x_2, x_3, ⋯, and x_n, the relation between y and x is multiplication of exponential involution.

$$y = k x_1^{\alpha_1} x_2^{\alpha_2} x_3^{\alpha_3} \cdots x_n^{\alpha_n} \tag{5-3}$$

where k—a numerical constant, for dimensional homogeneity, the both sides of the equation must be equal both numerically and dimensionally.

$$[y] = [k x_1^{\alpha_1} x_2^{\alpha_2} x_3^{\alpha_3} \cdots x_n^{\alpha_n}]$$

$$[L^a T^b M^c] = [L^{a_1} T^{b_1} M^{c_1}]^{\alpha_1} [L^{a_2} T^{b_2} M^{c_2}]^{\alpha_2} \cdots [L^{a_n} T^{b_n} M^{c_n}]^{\alpha_n}$$

The exponential of L, T and M of both sides must be equal respectively.

So
$$\left. \begin{array}{l} L: \quad a = a_1 \alpha_1 + a_2 \alpha_2 + \cdots + a_n \alpha_n \\ T: \quad b = b_1 \alpha_1 + b_2 \alpha_2 + \cdots + b_n \alpha_n \\ M: \quad c = c_1 \alpha_1 + c_2 \alpha_2 + \cdots + c_n \alpha_n \end{array} \right\} \tag{5-4}$$

The unknown exponentials of α_1, α_2, $\cdots$ and α_n will be obtained.

5.3 瑞利法和白金汉 π 定理

5.3.1 瑞利法

假如物理量 y 依赖于变量 x_1, x_2, x_3, $\cdots$, x_n 的变化，并且是幂指数乘积的函数关系，即

$$y = k x_1^{\alpha_1} x_2^{\alpha_2} x_3^{\alpha_3} \cdots x_n^{\alpha_n} \tag{5-3}$$

式中，k 是常数，那么根据量纲和谐原理，方程两边的数量及量纲均要求一致，即

$$[y] = [k x_1^{\alpha_1} x_2^{\alpha_2} x_3^{\alpha_3} \cdots x_n^{\alpha_n}]$$

$$[L^a T^b M^c] = [L^{a_1} T^{b_1} M^{c_1}]^{\alpha_1} [L^{a_2} T^{b_2} M^{c_2}]^{\alpha_2} \cdots [L^{a_n} T^{b_n} M^{c_n}]^{\alpha_n}$$

方程两边的 L、T、M 的指数对应分别相等。

因此
$$\left. \begin{array}{l} L: \quad a = a_1 \alpha_1 + a_2 \alpha_2 + \cdots + a_n \alpha_n \\ T: \quad b = b_1 \alpha_1 + b_2 \alpha_2 + \cdots + b_n \alpha_n \\ M: \quad c = c_1 \alpha_1 + c_2 \alpha_2 + \cdots + c_n \alpha_n \end{array} \right\} \tag{5-4}$$

未知数 α_1、$\alpha_2 \cdots \alpha_n$ 就可以求出。

【Sample Problem 5-1】 Under laminar flow condition, a fluid flows through a small equilateral triangle hole (side length is b, hole length is L). The volume flow rate Q is a function of dynamic viscosity μ, the pressure drop on unit length $\Delta p/L$ and b. Try to alter the relation into a dimensionless formula. If b doubles, what will be the change of Q?

Solution: When the number of unknown parameters is less than or equal to 4, we can obtain its accurate function with dimensional analysis method. It is called Rayleigh's method.

Write out exponential relationship as
$$Q = k \mu^\alpha \left(\frac{\Delta p}{L} \right)^\beta b^\gamma$$

where k—dimensionless coefficient to be solved.

Write out dimensional form as
$$[Q] = [\mu]^\alpha \left[\frac{\Delta p}{L} \right]^\beta [b]^\gamma$$

Express all dimensions of physical variables with basic dimension (L, T, M):

$$[L^3 T^{-1}] = [L^{-1} T^{-1} M]^\alpha [L^{-2} T^{-2} M]^\beta [L]^\gamma$$

Solve α, β, γ with dimensional homogeneity:

$$\left.\begin{array}{ll}\text{L:} & 3=-\alpha-2\beta+\gamma \\ \text{T:} & -1=-\alpha-2\beta \\ \text{M:} & 0=\alpha+\beta\end{array}\right\}$$

Then $\alpha=-1, \beta=1, \gamma=4$.

So we can obtain the following relationship:

$$Q=k\frac{b^4}{\mu}\left(\frac{\Delta p}{L}\right)$$

When b is doubled, the flow rate increases 15 times.

【例题 5-1】 在层流情况下，有液体流过一等边三角形截面的小孔（边长为 b，孔长为 L）。体积流量 Q 为动力黏度 μ、单位长度上的压降 $\Delta p/L$ 及 b 的函数。试将此关系写成无量纲式。若 b 加倍，流量有何变化？

解： 未知参数小于等于 4 个时，可直接用量纲分析法求得其准确的函数关系，这种方法即为瑞利法。

写出指数关系式

$$Q=k\mu^{\alpha}\left(\frac{\Delta p}{L}\right)^{\beta}b^{\gamma}$$

式中，k 为待求的无量纲系数。

写出量纲式

$$[Q]=[\mu]^{\alpha}\left[\frac{\Delta p}{L}\right]^{\beta}[b]^{\gamma}$$

以基本量纲（L，M，T）表示各物理量的量纲：

$$[L^3 T^{-1}]=[L^{-1}T^{-1}M]^{\alpha}[L^{-2}T^{-2}M]^{\beta}[L]^{\gamma}$$

根据量纲和谐原理求 α、β、γ

$$\left.\begin{array}{ll}\text{L:} & 3=-\alpha-2\beta+\gamma \\ \text{T:} & -1=-\alpha-2\beta \\ \text{M:} & 0=\alpha+\beta\end{array}\right\}$$

得 $\alpha=-1, \beta=1, \gamma=4$。

则 $Q=k\dfrac{b^4}{\mu}\left(\dfrac{\Delta p}{L}\right)$，当 b 增加 1 倍时，流量增加 15 倍。

5.3.2 Buckingham's π Theorem

Let $x_1, x_2, x_3, \cdots, x_n$ represent n dimensional variables, such as velocity, density and viscosity, which are involved in some physical phenomena. We can write the dimensionally homogeneous equation containing these variables as

$$f(x_1, x_2, x_3, \cdots, x_n)=0$$

In the equation, the number of independent variables is m. The others are the dependent variables and the number is $(n-m)$. The dimensions of each term in the equation are the same so we can rearrange this equation into the following equation.

$$F(\pi_1, \pi_2, \cdots, \pi_{n-m})=0$$

The parameters $\pi_1, \pi_2, \cdots, \pi_{n-m}$ are the dimensionless groups and the number of the groups is $(n-m)$. This method is known as the Buckingham's π Theorem and was reported firstly by E. Buckingham in 1915.

Suppose that there are three variables x_1, x_2, x_3 and three fundamental dimensions L,

T, M. The dimensions of the three variables will be represented as follows

$$\left.\begin{array}{l}\dim x_1 = [L^{a_1} T^{b_1} M^{c_1}] \\ \dim x_2 = [L^{a_2} T^{b_2} M^{c_2}] \\ \dim x_3 = [L^{a_3} T^{b_3} M^{c_3}]\end{array}\right\} \quad (5\text{-}5)$$

The matrix of powers of L, T and M can be formed as follows. If the three variables are independent each other, the value of the matrix should not be zero.

$$\begin{vmatrix} a_1 & b_1 & c_1 \\ a_2 & b_2 & c_2 \\ a_3 & b_3 & c_3 \end{vmatrix} \neq 0 \quad (5\text{-}6)$$

Now x_1, x_2, x_3 represent length, time and mass, respectively, so $a_1 = 1$, $b_2 = 1$, $c_3 = 1$.

$$\begin{vmatrix} 1 & 0 & 0 \\ 0 & 1 & 0 \\ 0 & 0 & 1 \end{vmatrix} = 1 \neq 0$$

The value of the matrix is one. x_1, x_2, x_3 are independent each other. If we choose the variables of which the dimensions belong to the three fundamental dimensions, it is easy to satisfy the independent condition of the variables.

The dimensionless parameter π comprises the independent variables and one of the rest variables. The dependent variables are chosen only one time. The dimensionless groups will be expressed as follows:

$$\left.\begin{array}{l}\pi_1 = x_1^{\alpha_1} x_2^{\beta_1} x_3^{\gamma_1} x_4 \\ \pi_2 = x_1^{\alpha_2} x_2^{\beta_2} x_3^{\gamma_2} x_5 \\ \cdots \\ \pi_{n-m} = x_1^{\alpha_{n-m}} x_2^{\beta_{n-m}} x_3^{\gamma_{n-m}} x_n\end{array}\right\} \quad (5\text{-}7)$$

According to dimensional homogeneity, the α_i, β_i, γ_i ($i = 1, 2, 3, \cdots, n-m$) can be confirmed.

We need to follow a series of six steps when applying the π theorem.

Step 1: Analyze the physical problem, consider the factors that are of influence, and list the n variables.

Step 2: Choose a dimensional system (L, T, M) and list the dimension of each variable. Find the number of fundamental dimensions involved in all the variables.

Step 3: Find the reduction number m. From the list of dimensional variables, select m of them to be so called independent variables.

Step 4: Determine $n-m$, the number of dimensionless π groups needed.

Step 5: To satisfy dimensional homogeneity, equate the exponents of each dimension on both sides of each π equation, and so solve for the exponents and the forms of the dimensionless groups.

Step 6: Rearrange the π groups as desired. The π theorem states that the πs are related, and may be expressed as $f_1(\pi_1, \pi_2, \pi_3, \cdots, \pi_{n-m}) = 0$ or as $\pi_1 = f_2(\pi_2, \pi_3, \cdots, \pi_{n-m})$.

5.3.2 白金汉 π 定理

$x_1, x_2, x_3, \cdots, x_n$ 表示 n 个有量纲的量，比如速度、密度及黏度等，它们包含在一些物理现象中，我们就可以写出含有这些变量的量纲和谐的物理方程。

$$f(x_1, x_2, x_3, \cdots x_n) = 0$$

其中有 m 个变量是独立的变量，其余的 $n-m$ 个变量是非独立变量，在方程中每一项的量纲是一致的，我们可以把方程写成另一种形式：

$$F(\pi_1, \pi_2, \cdots, \pi_{n-m}) = 0$$

$\pi_1、\pi_2 \cdots \pi_{n-m}$ 是 $n-m$ 个无量纲的量，这个方法就是白金汉 π 定理，是白金汉于 1915 年首先提出来的。

假设有三个变量 $x_1、x_2、x_3$，由三个基本量纲组成，它们的量纲形式可以表示成

$$\left. \begin{array}{l} \dim x_1 = [L^{a_1} T^{b_1} M^{c_1}] \\ \dim x_2 = [L^{a_2} T^{b_2} M^{c_2}] \\ \dim x_3 = [L^{a_3} T^{b_3} M^{c_3}] \end{array} \right\} \tag{5-5}$$

L、T、M 的指数可以表示成如下的行列式，如果这三个变量是相互独立的，那么行列式的值不等于零。

$$\begin{vmatrix} a_1 & b_1 & c_1 \\ a_2 & b_2 & c_2 \\ a_3 & b_3 & c_3 \end{vmatrix} \neq 0 \tag{5-6}$$

现在 $x_1、x_2、x_3$ 分别表示长度、时间及质量，则 $a_1 = 1, b_2 = 1, c_3 = 1$。

$$\begin{vmatrix} 1 & 0 & 0 \\ 0 & 1 & 0 \\ 0 & 0 & 1 \end{vmatrix} = 1 \neq 0$$

行列式的值等于 1，$x_1、x_2、x_3$ 是独立的，如果我们所选择的物理量的量纲分别属于此三种类型，则容易满足相互独立的条件。

无量纲的 π 项由所选的独立变量和其余变量中的一个变量组成，非独立变量只能选用一次。无量纲的数可以表示成如下形式：

$$\left. \begin{array}{l} \pi_1 = x_1^{\alpha_1} x_2^{\beta_1} x_3^{\gamma_1} x_4 \\ \pi_2 = x_1^{\alpha_2} x_2^{\beta_2} x_3^{\gamma_2} x_5 \\ \cdots \\ \pi_{n-m} = x_1^{\alpha_{n-m}} x_2^{\beta_{n-m}} x_3^{\gamma_{n-m}} x_n \end{array} \right\} \tag{5-7}$$

根据量纲和谐原理，确定 $\alpha_i、\beta_i、\gamma_i (i = 1, 2, 3, \cdots, n-m)$ 的值。

当运用 π 定理时，我们需要按照以下六个步骤来进行。

步骤 1：分析物理现象，考虑影响因子，列出 n 个变量。

步骤 2：选择 LTM 量纲制，列出每个变量的量纲，找出基本量纲及其数目。

步骤 3：确定要减少的变量的数目 m，从量纲列表中，选择 m 个独立变量。

步骤 4：确定 $n-m$，这就是无量纲的 π 的数量。

步骤 5：为了满足量纲和谐原理，π 表达式中每个量纲的指数应对应相等，且为无量纲的组合。

步骤 6：列出含 π 的方程 $f_1(\pi_1, \pi_2, \pi_3, \cdots, \pi_{n-m})=0$ 或者 $\pi_1=f_2(\pi_2, \pi_3, \cdots, \pi_{n-m})$。

【Sample Problem 5-2】 Based on actual observation, it is known that, in pipe flow, the pressure difference Δp caused by friction has relation to the following factors: pipe diameter d, average velocity v, dynamic viscosity μ, pipe length l, degree of roughness of pipe wall Δ, and fluid density ρ. Try to solve the friction head loss in pipe flow.

Solution: The relationship of the variables could be expressed as
$$f(d, v, \rho, \mu, l, \Delta, \Delta p)=0$$
The independent variables could be d, v and ρ.

Total number of variables, $n=7$.

Number of independent variables, $m=3$.

Number of dimensionless groups to be formed, $n-m=7-3=4$.

$$\pi_1 = d^{\alpha_1} v^{\beta_1} \rho^{\gamma_1} \mu$$
$$\pi_2 = d^{\alpha_2} v^{\beta_2} \rho^{\gamma_2} l$$
$$\pi_3 = d^{\alpha_3} v^{\beta_3} \rho^{\gamma_3} \Delta$$
$$\pi_4 = d^{\alpha_4} v^{\beta_4} \rho^{\gamma_4} \Delta p$$

For dimensional homogeneity
$$\dim \pi_1 = [L^0 T^0 M^0] = L^{\alpha_1} (LT^{-1})^{\beta_1} (L^{-3}M)^{\gamma_1} (L^{-1}T^{-1}M)$$

Equating powers of L, T and M
$$\left.\begin{array}{l} L: 0=\alpha_1+\beta_1-3\gamma_1-1 \\ T: 0=-\beta_1-1 \\ M: 0=\gamma_1+1 \end{array}\right\}$$

from which $\alpha_1=-1$, $\beta_1=-1$, $\gamma_1=-1$ and
$$\pi_1 = d^{-1} v^{-1} \rho^{-1} \mu = \frac{\mu}{vd\rho} = \frac{1}{Re}$$

where Re—Reynolds number.

Similarly, $\quad \dim \pi_2 = [L^0 T^0 M^0] = L^{\alpha_2} (LT^{-1})^{\beta_2} (L^{-3}M)^{\gamma_2} (L)$
$$\pi_2 = d^{-1} l = \frac{l}{d}$$

Similarly, $\quad \dim \pi_3 = [L^0 T^0 M^0] = L^{\alpha_3} (LT^{-1})^{\beta_3} (L^{-3}M)^{\gamma_3} (L)$
$$\pi_3 = d^{-1} \Delta = \frac{\Delta}{d}$$

Similarly, $\quad \dim \pi_4 = [L^0 T^0 M^0] = L^{\alpha_4} (LT^{-1})^{\beta_4} (L^{-3}M)^{\gamma_4} (L^{-1}T^{-2}M)$
$$\pi_4 = v^{-2} \rho^{-1} \Delta p = \frac{\Delta p}{v^2 \rho}$$

$$F\left(\frac{1}{Re}, \frac{l}{d}, \frac{\Delta}{d}, \frac{\Delta p}{v^2\rho}\right)=0$$

Or

$$\frac{\Delta P}{v^2\rho}=f\left(\frac{\mu}{dv\rho}, \frac{l}{d}, \frac{\Delta}{d}\right)$$

Because the friction head loss in pipe flow $h_f = \dfrac{\Delta p}{\rho g}$

so

$$h_f=\frac{\Delta P}{\rho g}=\frac{v^2}{g}f\left(\frac{1}{Re}, \frac{l}{d}, \frac{\Delta}{d}\right)$$

【例题 5-2】 根据实际观测知道，管中流动由于沿程摩擦而造成的压强差 Δp 与下列因素有关：管路直径 d，管中平均速度 v，流体动力黏度 μ，管路长度 l，管壁的粗糙度 Δ，流体的密度 ρ。试求水中流动的沿程水头损失。

解： 变量的关系可以表示为，

$$f(d, v, \rho, \mu, l, \Delta, \Delta p)=0$$

取 d、v、ρ 为独立变量，总变量数 $n=7$，独立变量数 $m=3$，无量纲量有 $n-m=7-3=4$ 个。

$$\pi_1 = d^{\alpha_1} v^{\beta_1} \rho^{\gamma_1} \mu$$
$$\pi_2 = d^{\alpha_2} v^{\beta_2} \rho^{\gamma_2} l$$
$$\pi_3 = d^{\alpha_3} v^{\beta_3} \rho^{\gamma_3} \Delta$$
$$\pi_4 = d^{\alpha_4} v^{\beta_4} \rho^{\gamma_4} \Delta p$$

根据量纲和谐原理

$$\dim \pi_1 = [L^0 T^0 M^0] = L^{\alpha_1} (LT^{-1})^{\beta_1} (L^{-3}M)^{\gamma_1} (L^{-1}T^{-1}M)$$

由 L、T 和 M 的指数对应相等，可得

$$\left.\begin{array}{l} L: 0=\alpha_1+\beta_1-3\gamma_1-1 \\ T: 0=-\beta_1-1 \\ M: 0=\gamma_1+1 \end{array}\right\}$$

解得 $\alpha_1=-1$、$\beta_1=-1$、$\gamma_1=-1$ 及 $\pi_1=d^{-1}v^{-1}\rho^{-1}\mu=\dfrac{\mu}{vd\rho}=\dfrac{1}{Re}$，$Re$ 为雷诺数。

同样，

$$\dim \pi_2 = [L^0 T^0 M^0] = L^{\alpha_2} (LT^{-1})^{\beta_2} (L^{-3}M)^{\gamma_2} (L)$$

$$\pi_2 = d^{-1} l = \frac{l}{d}$$

同样，

$$\dim \pi_3 = [L^0 T^0 M^0] = L^{\alpha_3} (LT^{-1})^{\beta_3} (L^{-3}M)^{\gamma_3} (L)$$

$$\pi_3 = d^{-1} \Delta = \frac{\Delta}{d}$$

同样，

$$\dim \pi_4 = [L^0 T^0 M^0] = L^{\alpha_4} (LT^{-1})^{\beta_4} (L^{-3}M)^{\gamma_4} (L^{-1}T^{-2}M)$$

$$\pi_4 = v^{-2} \rho^{-1} \Delta p = \frac{\Delta p}{v^2 \rho}$$

$$F\left(\frac{1}{Re}, \frac{l}{d}, \frac{\Delta}{d}, \frac{\Delta p}{v^2\rho}\right)=0$$

或

$$\frac{\Delta P}{v^2\rho}=f\left(\frac{\mu}{dv\rho}, \frac{l}{d}, \frac{\Delta}{d}\right)$$

因为管中流动的水头损失

$$h_f = \frac{\Delta p}{\rho g}$$

所以

$$h_f = \frac{\Delta P}{\rho g} = \frac{v^2}{g} f\left(\frac{1}{Re}, \frac{l}{d}, \frac{\Delta}{d}\right)$$

5.4 Similarity Principle

When we design or manufacture some complex and huge hydraulic machines, build hydraulic engineering as well as research some complicated hydraulic phenomena, a model with reduced size is often designed and made according to similarity principle. Through carrying out simulation experiment and observing the flow situation of model, we deduce the flow situation of real object and relative data. The basic theory which analyzes and studies the similarity relationship between model and real object is called similarity principle. Real object is also called prototype.

5.4 相似原理

当设计制造某些复杂而庞大的水力机械、建造水利工程以及研究某些复杂的水力现象时，往往要根据相似原理，设计制造尺寸缩小了的模型进行模拟试验，通过对模型流动状况的观测来推断实物的流动状况及有关数据。分析研究模型和实物间的相似关系的基本理论称为相似原理。实物又称为原型。

5.4.1 Geometric Similarity

When the corresponding geometrical linear dimensions of prototype are in proportion to those of model and corresponding geometrical angles are equivalent, it is called geometric similarity.

Model parameter adds a subscript m and prototype parameter adds a subscript p to express.

Linear scale
$$\lambda_l = \frac{l_p}{l_m} \tag{5-8}$$

Surface scale
$$\lambda_A = \frac{A_p}{A_m} = \frac{l_p^2}{l_m^2} = \lambda_l^2 \tag{5-9}$$

Volume scale
$$\lambda_V = \frac{V_p}{V_m} = \frac{l_p^3}{l_m^3} = \lambda_l^3 \tag{5-10}$$

5.4.1 几何相似

原型与模型中对应的几何线性尺寸成比例，对应的几何角度相等，称为几何相似。

模型参数以下标 m、原型参数以下标 p 表示。

线性比尺为
$$\lambda_l = \frac{l_p}{l_m} \tag{5-8}$$

面积比尺为
$$\lambda_A = \frac{A_p}{A_m} = \frac{l_p^2}{l_m^2} = \lambda_l^2 \tag{5-9}$$

体积比尺为
$$\lambda_V = \frac{V_p}{V_m} = \frac{l_p^3}{l_m^3} = \lambda_l^3 \tag{5-10}$$

5.4.2 Kinematic Similarity

When the corresponding parameters of motion of prototype such as velocity, acceleration are consistent in direction and in proportion to magnitude to those of model, it is called kinematic similarity.

Time scale
$$\lambda_t = \frac{t_p}{t_m} \tag{5-11}$$

Velocity scale
$$\lambda_v = \frac{v_p}{v_m} = \frac{\frac{l_p}{t_p}}{\frac{l_m}{t_m}} = \frac{\lambda_l}{\lambda_t} \tag{5-12}$$

Acceleration scale
$$\lambda_a = \frac{a_p}{a_m} = \frac{\frac{l_p}{t_p^2}}{\frac{l_m}{t_m^2}} = \frac{\lambda_l}{\lambda_t^2} \tag{5-13}$$

5.4.2 运动相似

原型与模型中对应的运动参数如速度、加速度方向一致，大小成比例，称为运动相似。

时间比尺为
$$\lambda_t = \frac{t_p}{t_m} \tag{5-11}$$

速度比尺为
$$\lambda_v = \frac{v_p}{v_m} = \frac{\frac{l_p}{t_p}}{\frac{l_m}{t_m}} = \frac{\lambda_l}{\lambda_t} \tag{5-12}$$

加速度比尺为
$$\lambda_a = \frac{a_p}{a_m} = \frac{\frac{l_p}{t_p^2}}{\frac{l_m}{t_m^2}} = \frac{\lambda_l}{\lambda_t^2} \tag{5-13}$$

5.4.3 Dynamic Similarity

When the forces of corresponding points of prototype are consistent in direction and in proportion to magnitude to those of model, it is called dynamic similarity.

Density scale
$$\lambda_\rho = \frac{\rho_p}{\rho_m} \tag{5-14}$$

Mass scale
$$\lambda_m = \frac{m_p}{m_m} = \frac{\rho_p V_p}{\rho_m V_m} = \lambda_\rho \lambda_l^3 \tag{5-15}$$

Force scale
$$\lambda_F = \frac{F_p}{F_m} = \frac{m_p a_p}{m_m a_m} = \lambda_m \lambda_a = \lambda_\rho \lambda_v^2 \lambda_l^2 \tag{5-16}$$

Unit body force scale
$$\lambda_g = \frac{g_p}{g_m} \tag{5-17}$$

According to Equation (5-16), we obtain $\dfrac{F_p}{F_m}=\dfrac{\rho_p l_p^2 v_p^2}{\rho_m l_m^2 v_m^2}$

namely
$$\dfrac{F_p}{\rho_p l_p^2 v_p^2}=\dfrac{F_m}{\rho_m l_m^2 v_m^2} \tag{5-18}$$

In this equation, $\dfrac{F}{\rho l^2 v^2}$, a dimensionless number, is called Newton number and denoted by Ne.

So Equation (5-18) changes into
$$Ne_p = Ne_m \tag{5-19}$$

That is to say, if two geometrically similar flows are dynamically similar, then their Newton numbers must be equal; whereas, two geometrically similar flows, whose Newton numbers are equal, must be dynamically similar. So geometric similarity is only necessary condition of similarity, and kinematic similarity and dynamic similarity are necessary and sufficient conditions of similarity.

5.4.3 动力相似

原型与模型中对应点处受力方向相同、大小成比例，称为动力相似。

密度比尺为
$$\lambda_\rho = \dfrac{\rho_p}{\rho_m} \tag{5-14}$$

质量比尺为
$$\lambda_m = \dfrac{m_p}{m_m} = \dfrac{\rho_p V_p}{\rho_m V_m} = \lambda_\rho \lambda_l^3 \tag{5-15}$$

力的比尺为
$$\lambda_F = \dfrac{F_p}{F_m} = \dfrac{m_p a_p}{m_m a_m} = \lambda_m \lambda_a = \lambda_\rho \lambda_v^2 \lambda_l^2 \tag{5-16}$$

单位质量力比尺为
$$\lambda_g = \dfrac{g_p}{g_m} \tag{5-17}$$

据式(5-16)可得
$$\dfrac{F_p}{F_m}=\dfrac{\rho_p l_p^2 v_p^2}{\rho_m l_m^2 v_m^2}$$

即
$$\dfrac{F_p}{\rho_p l_p^2 v_p^2}=\dfrac{F_m}{\rho_m l_m^2 v_m^2} \tag{5-18}$$

式中，$\dfrac{F}{\rho l^2 v^2}$为一个无量纲量，称为牛顿数，以 Ne 表示。故式(5-18)变为
$$Ne_p = Ne_m \tag{5-19}$$

就是说，两个几何相似的流动，如果动力相似，则牛顿数必相等；反之，牛顿数相等的两个几何相似的流动，必然是动力相似的。故几何相似仅是相似的必要条件，而运动相似和动力相似才是相似的充要条件。

5.5 Similarity Criterion

5.5.1 Reynolds Number

Internal friction caused by viscosity is

$$T = \mu A \frac{du}{dy}$$

That can be expressed from dimension

$$[T] = [\mu][L^2]\left[\frac{v}{L}\right] = [\mu l v]$$

From Equation (5-18) we know the dimension of inertial force is $[F] = [\rho l^2 v^2]$, if replace F with T, then

$$\frac{\mu_p l_p v_p}{\rho_p l_p^2 v_p^2} = \frac{\mu_m l_m v_m}{\rho_m l_m^2 v_m^2}$$

For $\frac{\mu}{\rho} = \nu$, so simplifying the formula above we get:

$$\frac{v_p l_p}{\nu_p} = \frac{v_m l_m}{\nu_m} \tag{5-20}$$

In this formula, $\frac{vl}{\nu} = Re$, and Re is called Reynolds number. Its physical meaning is the ratio of inertial force to viscosity force.

5.5 相似准则

5.5.1 雷诺数

黏性引起的内摩擦力为

$$T = \mu A \frac{du}{dy}$$

根据量纲可写为

$$[T] = [\mu][L^2]\left[\frac{v}{L}\right] = [\mu l v]$$

由式(5-18)知惯性力的量纲为 $[F] = [\rho l^2 v^2]$，如用 T 替换 F，则

$$\frac{\mu_p l_p v_p}{\rho_p l_p^2 v_p^2} = \frac{\mu_m l_m v_m}{\rho_m l_m^2 v_m^2}$$

因 $\frac{\mu}{\rho} = \nu$，故简化上式得

$$\frac{v_p l_p}{\nu_p} = \frac{v_m l_m}{\nu_m} \tag{5-20}$$

式中，$\frac{vl}{\nu} = Re$，称为雷诺数（Reynolds number），其物理意义为惯性力与黏性力之比。

5.5.2 Froude Number

In a liquid with a free surface, the predominant force is gravity: $G = mg = \rho g V$. On dimension,

$$[G] = [\rho][g][L^3] = [\rho g l^3]$$

Substitution of G for F in Equation (5-18) yields

$$\frac{\rho_p g_p l_p^3}{\rho_p l_p^2 v_p^2} = \frac{\rho_m g_m l_m^3}{\rho_m l_m^2 v_m^2}$$

Simplifying

$$\frac{v_p^2}{g_p l_p} = \frac{v_m^2}{g_m l_m} \tag{5-21}$$

In this formula, $\frac{v}{\sqrt{gl}} = Fr$, and Fr is called Froude number. Its physical meaning is the ratio of inertial force to gravity.

5.5.2 弗劳德数

在具有自由表面的液流中，起主要作用的为重力 $G = mg = \rho g V$，在量纲上为

$$[G] = [\rho][g][L^3] = [\rho g l^3]$$

用 G 代替式(5-18) 中的 F，则

$$\frac{\rho_p g_p l_p^3}{\rho_p l_p^2 v_p^2} = \frac{\rho_m g_m l_m^3}{\rho_m l_m^2 v_m^2}$$

简化后得

$$\frac{v_p^2}{g_p l_p} = \frac{v_m^2}{g_m l_m} \tag{5-21}$$

式中，$\frac{v}{\sqrt{gl}} = Fr$，称为弗劳德数（Froude number），其物理意义为惯性力与重力之比。

5.5.3 Euler Number

When we study the distribution of press or pressure on the surface of objects submerged in liquids, the major force is press $F = pA$. On dimension,

$$[F] = [p][A] = [pl^2]$$

Substitution of it for F in Equation (5-18) yields

$$\frac{p_p l_p^2}{\rho_p l_p^2 v_p^2} = \frac{p_m l_m^2}{\rho_m l_m^2 v_m^2}$$

Namely

$$\frac{p_p}{\rho_p v_p^2} = \frac{p_m}{\rho_m v_m^2} \tag{5-22}$$

where $\frac{p}{\rho v^2} = Eu$. Eu is called Euler number. Its physical meaning is the ratio of press to inertial force.

If two flows are dynamically similar, they must have the same Froude number, Euler number and Reynolds number, that is

$$\left.\begin{array}{l} Fr_p = Fr_m \\ Eu_p = Eu_m \\ Re_p = Re_m \end{array}\right\} \tag{5-23}$$

It could be known as mechanical similarity criterion for the steady flow of incompressible fluid.

5.5.3 欧拉数

研究淹没在流体中的物体表面上的压力或压强分布时，起主要作用的力为压力 $F = pA$。

在量纲上为
$$[F] = [p][A] = [pl^2]$$
用其代替式(5-18)中的 F，则
$$\frac{p_p l_p^2}{\rho_p l_p^2 v_p^2} = \frac{p_m l_m^2}{\rho_m l_m^2 v_m^2}$$
即
$$\frac{p_p}{\rho_p v_p^2} = \frac{p_m}{\rho_m v_m^2} \tag{5-22}$$

式中，$\frac{p}{\rho v^2} = Eu$，称为欧拉数（Euler number），其物理意义为压力与惯性力之比。

如果两个流动动力学相似，则它们的弗劳德数、欧拉数、雷诺数必各自相等，于是
$$\left. \begin{array}{l} Fr_p = Fr_m \\ Eu_p = Eu_m \\ Re_p = Re_m \end{array} \right\} \tag{5-23}$$

这称为不可压缩流体定常流动的力学相似准则。

5.5.4 Mach Number

When the fluid compressibility is thought about, the predominant force is elastic force, $F = EA$. On dimension,
$$[F] = [E][A] = [El^2]$$
Substitute it into Equation (5-18), and then
$$\frac{E_p l_p^2}{\rho_p l_p^2 v_p^2} = \frac{E_m l_m^2}{\rho_m l_m^2 v_m^2}$$
Namely
$$\frac{E_p}{\rho_p v_p^2} = \frac{E_m}{\rho_m v_m^2}$$

For compressible fluid, velocity of sound is $a = \sqrt{\frac{E}{\rho}}$, so $\frac{\rho}{E} = \frac{1}{a^2}$.

Substituting it into the formula above, we obtain $\frac{a_p}{v_p} = \frac{a_m}{v_m}$. Its reciprocal $\frac{v}{a} = Ma$, and Ma is called Mach number. Its physical meaning is the ratio of inertial force to elastic force.

5.5.4 马赫数

当考虑流体压缩性时，弹性力起主要作用，$F = EA$。在量纲上
$$[F] = [E][A] = [El^2]$$
代入式(5-18)中的 F 时，则
$$\frac{E_p l_p^2}{\rho_p l_p^2 v_p^2} = \frac{E_m l_m^2}{\rho_m l_m^2 v_m^2}$$
即
$$\frac{E_p}{\rho_p v_p^2} = \frac{E_m}{\rho_m v_m^2}$$

对可压缩流体，声速 $a = \sqrt{\frac{E}{\rho}}$，因此 $\frac{\rho}{E} = \frac{1}{a^2}$。

代入上式得 $\dfrac{a_p}{v_p}=\dfrac{a_m}{v_m}$，其倒数 $\dfrac{v}{a}=Ma$，称为马赫数，其物理意义为惯性力与弹性力之比。

5.5.5 Weber Number

Substituting surface tension $F=\sigma l$ into Equation (5-18), we can educe
$$\frac{\rho_p l_p v_p^2}{\sigma_p}=\frac{\rho_m l_m v_m^2}{\sigma_m}$$

Let the ratio $\dfrac{\rho l v^2}{\sigma}=We$, and We is called Weber number. Its physical meaning is the ratio of inertial force to surface tension.

5.5.5 韦伯数

将表面张力 $F=\sigma l$ 代入式(5-18)中，可导出
$$\frac{\rho_p l_p v_p^2}{\sigma_p}=\frac{\rho_m l_m v_m^2}{\sigma_m}$$

令该比值 $\dfrac{\rho l v^2}{\sigma}=We$，称为韦伯数，其物理意义为惯性力与表面张力之比。

5.6 Model Experiment

Experiment is the basis of theory development as well as the yardstick to prove theories. Sometimes science and technology problems can't be solved without the cooperation of experiments. There are mainly two kinds of experiments in engineering fluid mechanics. One is model experiment of engineering. The purpose is to forecast the flow situation of large-scale machine and hydraulic construction which are being built. The other one is exploratory observing experiment. For the purpose of searching unknown flow laws, the theoretical foundations to direct these experiments are similarity principle and dimension analysis.

In the same physical phenomena, two similarity numbers can't often meet similarity relationship at the same time. For example, Reynolds number and Froude number are not easy to meet the conditions at the same time.

To make Reynolds numbers equal:
$$\frac{v_p l_p}{\nu_p}=\frac{v_m l_m}{\nu_m}, \text{ namely, } \frac{\lambda_v \lambda_l}{\lambda_\nu}=1$$

To make Froude numbers equal:
$$\frac{v_p}{\sqrt{g_p l_p}}=\frac{v_m}{\sqrt{g_m l_m}}, \text{ namely, } \frac{\lambda_v}{\sqrt{\lambda_g \lambda_l}}=1$$

If the two similarity numbers can meet the similarity relationship, it is necessary that the following equation exists:
$$\frac{\lambda_v \lambda_l}{\lambda_\nu}=\frac{\lambda_v}{\sqrt{\lambda_g \lambda_l}}$$

where $\lambda_g=1$. So we can obtain

$$\frac{\lambda_v}{\lambda_l}=\lambda_l^{\frac{1}{2}}, \text{ namely } \lambda_\nu=\lambda_l^{\frac{3}{2}}$$

It is hard or unable to realize on technique. In fact, we often analyze the flow problems deeply to find out the main acting force which influences flow problems, and then the main force similarity is met and the other minor force similarities are ignored.

For the pressure flow in the pipe, submerged body ambient flow, etc., as long as the Reynolds number of the flow is not too large, commonly its similarity condition depends on Reynolds criterion. While the wave motion caused by ship movement, water flow in open channel, water flow around pier, and jet from small hole in container wall, etc., are mainly influenced by gravity, and the similarity condition must ensure that Froude numbers are equal.

5.6 模型试验

实验既是发展理论的依据，又是检验理论的准绳，解决科技问题往往离不开实验手段的配合。工程流体力学中的实验主要有两类：一类是工程性的模型试验，目的在于预测即将建造的大型机械或水工结构上的流动情况；另一类是探索性的观察实验，目的在于寻找未知的流动规律，指导这些实验的理论基础就是相似原理和量纲分析。

两个相似准数在同一个物理现象中往往不能同时满足相似关系，例如雷诺数和弗劳德数就不易同时满足。

欲使雷诺数相等，将有

$$\frac{v_p l_p}{\nu_p}=\frac{v_m l_m}{\nu_m}, \quad 即 \quad \frac{\lambda_v \lambda_l}{\lambda_\nu}=1$$

欲使弗劳德数相等，将有

$$\frac{v_p}{\sqrt{g_p l_p}}=\frac{v_m}{\sqrt{g_m l_m}}, \quad 即 \quad \frac{\lambda_v}{\sqrt{\lambda_g \lambda_l}}=1$$

若想同时满足雷诺数相等和弗劳德数相等，必须有

$$\frac{\lambda_v \lambda_l}{\lambda_\nu}=\frac{\lambda_v}{\sqrt{\lambda_g \lambda_l}}$$

而 $\lambda_g=1$，则

$$\frac{\lambda_v}{\lambda_l}=\lambda_l^{\frac{1}{2}}, \quad 即 \quad \lambda_\nu=\lambda_l^{\frac{3}{2}}$$

这在技术上很难甚至不可能做到。实际上，常常要对所研究的流动问题做深入的分析，找出影响流动问题的主要作用力，满足这个主要力的相似而忽略其他次要力的相似。

对于管中的有压流动及潜体绕流等，只要流动的雷诺数不是特别大，一般其相似条件依赖于雷诺准则，而行船引起的波浪运动、明渠水流、绕桥墩的水流、容器壁小孔射流等主要受重力影响，相似条件要保证弗劳德数相等。

Exercises 习题

5-1 Use dimensional analysis to arrange the following groups into dimensionless parameters: (a) τ, V, ρ; (b) V, L, ρ, σ. Use the MLT system.

5-1 利用量纲分析法，把下列每组参数表示成无量纲量：(a) τ, V, ρ；(b) V, L, ρ, σ。使

用 MLT 制。

5-2 Use dimensional analysis to arrange the following groups into dimensionless parameters：(a) Δp, V, γ, g；(b) F, ρ, L, V. Use the MLT system.

5-2 利用量纲分析法，把下列每组参数表示成无量纲量：(a) Δp, V, γ, g；(b) F, ρ, L, V。使用 MLT 制。

5-3 Use dimensional analysis to derive an expression for the power developed by an engine in terms of the torque T and rotative speed ω.

5-3 利用量纲分析法求发动机产生的功的表达式，用扭矩 T 和旋转角速度 ω 表示。

5-4 Derive an expression for the shear stress at the pipe wall when an incompressible fluid flows through a pipe under pressure. Use dimensional analysis with the following significant parameters：pipe diameter D, flow velocity v, viscosity μ and density ρ of the fluid.

5-4 当不可压缩的液体在压力作用下流经水管时，求对管壁的剪切应力的表达式。根据下列重要的参数用量纲分析法分析：管径 D、液体的流速 v、黏度 μ、密度 ρ。

5-5 The movement of a flying object in stationary air is the same as the flow of airflow around fixed object. According to measurement, in circulating flow the flow resistance D has relation to transect linear dimension l, average velocity v, air density ρ and dynamic viscosity μ. Try to analyze the resistance formula of profile flow.

5-5 飞行物体在静止空气中运动相当于气流绕固定物体流动。根据测定，绕流时流体所受到的阻力 D 与物体横断面线性尺寸 l、气流平均流速 v、气体密度 ρ、动力黏度 μ 有关，试分析绕流阻力公式。

5-6 The acceleration, a, sometimes used in compressible flow theory, is a dimensionless combination of acceleration of gravity g, dynamic viscosity μ, density ρ, and bulk modulus B. If a is inversely proportional to density, find its expression.

5-6 加速度 a 往往用于可压缩流体理论，是一个由重力加速度 g、动力黏度 μ、密度 ρ 及体积弹性模量 B 组成的无量纲量。如果 a 与密度成反比，写出其表达式。

5-7 The wall shear stress τ_w in a boundary layer is assumed to be a function of stream velocity U, boundary layer thickness δ, local turbulence velocity u', density ρ, and local pressure gradient dp/dx. Using (ρ, U, δ) as repeating variables, rewrite this relationship as a dimensionless function.

5-7 假定边界层壁面剪应力 τ_w 是速度 U、边界层厚度 δ、局部湍流速度 u'、密度 ρ 和局部压力梯度 dp/dx 的函数。使用 (ρ, U, δ) 为重复的变量，用一个无量纲的函数关系式重新表示这种关系。

5-8 When coasting by inertia, the angular velocity Ω of a windmill is found to be a function of the windmill diameter D, the wind velocity V, the air density ρ, the windmill height H as compared to the atmospheric boundary layer height L, and the number of blades N：

$$\Omega = f\left(D, V, \rho, \frac{H}{L}, N\right)$$

Viscosity effects are negligible. Find appropriate π groups for this problem and rewrite the function in dimensionless form.

5-8 当惯性滑行时风车的角速度 Ω 被认为与其直径 D、风速 V、空气密度 ρ、风车高度 H 与大气边界层高度 L 之比及叶片数量 N 存在函数关系：

$$\Omega = f\left(D, V, \rho, \frac{H}{L}, N\right)$$

黏性的影响可以忽略不计。找到对应这一问题适当的 π 组合，用一个无量纲形式重新写出该函数关系式。

5-9 Use the length scale $\lambda_1 = 10$ model to test the aerodynamic characteristics of shells. Given that the shell's flying velocity is 1000m/s, air temperature is 40℃, aerodynamic viscosity is 19.2×10^{-6} Pa·s; air temperature of model is 10℃, aerodynamic viscosity of model is 17.8×10^{-6} Pa·s. Try to obtain the wind speed and wind pressure of model which meets elastic force similarity and viscosity force similarity.

5-9 用长度比例尺 $\lambda_1 = 10$ 的模型试验炮弹的空气动力特性。已知炮弹的飞行速度为 1000m/s，空气温度为 40℃，空气动力黏度为 19.2×10^{-6} Pa·s；模型的空气温度为 10℃，空气动力黏度为 17.8×10^{-6} Pa·s。试求同时满足弹性力相似和黏性力相似的模型的风速和风压。

5-10 A model spillway has a flow of 100L/s per meter of width. What is the actual flow for the prototype spillway if the model scale is 1∶20?

5-10 一模型溢流坝单宽流量为 100L/s，如果模型比尺为 1∶20，那么原型溢流坝的实际流量为多少？

5-11 Use dimensional analysis to arrange velocity v, length l and acceleration of gravity g into dimensionless parameters. The correct answer is _____ .

5-11 速度 v、长度 l、重力加速度 g 的无量纲组合是_____。

(A) $\dfrac{lv}{g}$ (B) $\dfrac{v}{gl}$ (C) $\dfrac{v^2}{gl}$ (D) $\dfrac{l}{gv}$

5-12 Use dimensional analysis to arrange velocity v, density ρ and pressure p into dimensionless parameters. The correct answer is _____ .

5-12 速度 v、密度 ρ、压强 p 的无量纲组合是_____。

(A) $\dfrac{\rho p}{v}$ (B) $\dfrac{\rho v}{p}$ (C) $\dfrac{p v^2}{\rho}$ (D) $\dfrac{p}{\rho v^2}$

5-13 Use dimensional analysis to arrange velocity v, length l and time t into dimensionless parameters. The correct answer is _____ .

5-13 速度 v、长度 l、时间 t 的无量纲组合是_____。

(A) $\dfrac{l}{vt}$ (B) $\dfrac{l}{vt^2}$ (C) $\dfrac{t}{vl}$ (D) $\dfrac{v}{lt}$

5-14 Use dimensional analysis to arrange force F, density ρ, length l and velocity v into dimensionless parameters. The correct answer is _____ .

5-14 力 F、密度 ρ、长度 l、速度 v 的无量纲组合是_____。

(A) $\dfrac{F}{\rho l v}$ (B) $\dfrac{F}{\rho l^2 v}$ (C) $\dfrac{F}{\rho l v^2}$ (D) $\dfrac{F}{\rho l^2 v^2}$

5-15 Use dimensional analysis to arrange dynamic viscosity μ, velocity v, length l and density ρ into dimensionless parameters. The correct answer is _____ .

5-15 动力黏度 μ、速度 v、长度 l、密度 ρ 的无量纲组合是_____。

(A) $\dfrac{\mu}{\rho l v}$ (B) $\dfrac{\mu}{\rho l^2 v}$ (C) $\dfrac{\mu}{\rho l v^2}$ (D) $\dfrac{\mu}{\rho l^2 v^2}$

Chapter 6　Flow Resistance and Energy Loss
第 6 章　流动阻力和能量损失

6.1　Laminar and Turbulent Flows

Some basic concepts about the rules of incompressible fluid motion in pipe are always adaptable to the detour flow and open channel flow. The questions involved in pipe flow include flow type, velocity distribution, beginning segment, the calculation of flux and pressure and energy loss, etc.

6.1　层流和紊流

管中不可压缩流体的运动规律，其中有许多基本概念对于绕流或明渠流动也是适用的。管流所涉及的问题包括流动状态、速度分布、起始段、流量和压强的计算、能量损失等。

6.1.1　Reynolds Experiment

Reynolds number reflects the ratio of inertial force to viscosity force. If Reynolds numbers are different, the ratios of the two kinds of forces are different. Therefore, two kinds of flow types, of which the interior structure and motion character are completely different, are produced.

As shown in Figure 6-1, the experimental apparatus is composed mainly of the container of constant water level and glass pipe. The entrance of glass pipe is connected by slippy bell mouth and the flux in pipe is adjusted by valve C. In the small container there is colored liquid whose density is close to water. The colored liquid flows into glass pipe via the tubule to demonstrate the flow type.

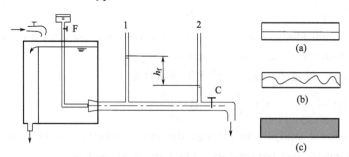

Figure 6-1　Reynolds Experimental Apparatus
图 6-1　雷诺实验装置

① As shown in Figure 6-1 (a), when velocity of fluid in pipe is less, the colored liquid in pipe presents a thin and well defined straight stream tube. This shows the flow is stable, and this kind of flow type is called laminar flow.

② As shown in Figure 6-1 (b), when the valve C enlarges gradually and the velocity of flow in pipe reaches a certain critical value, the colored liquid begins to swing.

③ Continue to augment the velocity of flow, as shown in Figure 6-1 (c), the colored liquid mixes with the clean water around it quickly. It is indicated the motion trace of fluid particle was extremely irregular and the fluids mixed with each other intensely. This kind of flow type is called turbulent flow.

6.1.1 雷诺实验

雷诺数反映惯性力和黏性力的比值大小。雷诺数不同,这两种力的比值也不同,由此产生内部结构和运动性质完全不同的两种流动状态。

雷诺实验的装置如图 6-1 所示,主要由恒水位水箱和玻璃管等组成。玻璃管入口部分用光滑喇叭口连接,管中的流量用阀门 C 调节,小瓶子内盛有与水的密度相近的有色液体,经细管流入玻璃管,用以演示水流状态。观察到以下现象。

① 当玻璃管内流速较小时,管内有色液体呈一细股界线分明的直线流束,见图 6-1(a),表明流动稳定,这种流动状态称为层流。

② 当阀门 C 逐渐开大使管中流速达到某一临界值时,有色液体开始出现摆动,见图 6-1 (b)。

③ 继续增大流速,有色液体迅速与周围清水相掺混,如图 6-1 (c)所示,表明流体质点的运动轨迹是极不规则的,流体互相剧烈掺混,这种运动状态称为紊流或湍流。

6.1.2 The Criterion of Laminar FLows and Turbulent Flows

As the pipe diameter d and fluid kinematic viscosity ν are constant, the mean velocity when laminar flow just changes to turbulent flow is constant too. This velocity is called supercritical velocity which is expressed by v'_{cr}. Similarly, the mean velocity when turbulent flow just changes to laminar flow is also constant, and this velocity is called lower critical velocity which is expressed by v_{cr}. There is a relation: $v'_{cr} > v_{cr}$.

The flow type has relations with not only velocity of flow v but also pipe diameter d, fluid density ρ and kinematic viscosity ν. According to the dimensional analysis in Chapter 5 we can arrange the four parameters into a dimensionless number, Reynolds number.

$$Re = \frac{vd\rho}{\mu} = \frac{vd}{\nu} \tag{6-1}$$

The Reynolds number corresponding to the critical velocity of flow v_{cr} is called critical Reynolds number, marked as Re_{cr}.

As pipe diameter and fluid medium are different, v_{cr} are different. But the experiments show that Re_{cr} keeps in a definite range basically, namely $Re_{cr} \approx 2300$.

For the flow in a round pipe, the criterions of the flow type are as follows:

When $Re < Re_{cr}$, it belongs to laminar flow;

When $Re > Re_{cr}$, it belongs to turbulent flow.

If hydraulic radius is R, Reynolds number can be denoted as

$$Re = \frac{vR}{\nu} \tag{6-2}$$

$$R = \frac{A}{\chi} \tag{6-3}$$

where A —area of cross-section;

χ —wetted perimeter (perimeter of solid boundary contacted with liquid of cross-section).

If hydraulic radius R is substituted into the formula of Reynolds number as a characteristic length, the critical Reynolds number is 580.

6.1.2 层流和紊流的判别标准

当管径 d 及流体运动黏度 ν 一定时，从层流变紊流时的平均速度也是一定的，此速度称为上临界速度，以 v'_{cr} 表示；从紊流变层流时的平均速度也是一定的，此速度称为下临界速度，以 v_{cr} 表示。$v'_{cr} > v_{cr}$。

流动状态不仅与流速 v 有关，还与管径 d、流体密度 ρ 和运动黏度 ν 有关。根据第5章的量纲分析方法可以将上述四个参数组合成一个无量纲数——雷诺数。

$$Re = \frac{vd\rho}{\mu} = \frac{vd}{\nu} \tag{6-1}$$

对应于临界流速 v_{cr} 的雷诺数称为临界雷诺数，记作 Re_{cr}。虽然当管径或流体介质不同时，v_{cr} 不同，但实验表明 Re_{cr} 基本上保持在一个确定的范围，即 $Re_{cr} \approx 2300$。

对圆管流动，流态的判别条件是：

$$Re < Re_{cr} \qquad 属于层流$$
$$Re > Re_{cr} \qquad 属于紊流$$

若水力半径为 R，此时雷诺数记为

$$Re = \frac{vR}{\nu} \tag{6-2}$$

$$R = \frac{A}{\chi} \tag{6-3}$$

式中，A 为过水断面面积；χ 为湿周（断面中固体边界与流体相接触部分的周长）。若雷诺数中的特征长度用水力半径 R 表示，则临界雷诺数的数值为 580。

【Sample Problem 6-1】 There is a tap water pipeline. Diameter $d = 100$mm, velocity $v = 1.0$m/s, and the water temperature is 10℃. Try to distinguish the type of flow in the pipe.

Solution: At 10℃, kinematic viscosity $\nu = 0.0131$cm^2/s, so the Reynolds number of flow in pipe is

$$Re = \frac{vd}{\nu} = \frac{100 \times 10}{0.0131} = 76335.9$$

$$Re_{cr} = 2300, \quad Re > Re_{cr}$$

This shows that the flow in pipe is turbulent.

【例题 6-1】 某段自来水管，其管径 $d = 100$mm，管中流速 $v = 1.0$m/s，水的温度为 10℃，试判别管中的水流形态。

解：在温度为 10℃时，水的运动黏度 $\nu = 0.0131$cm^2/s，管中水流的雷诺数

$$Re = \frac{vd}{\nu} = \frac{100 \times 10}{0.0131} = 76335.9$$

$$Re_{cr} = 2300, \quad Re > Re_{cr}$$

因此，管中水流处于紊流形态。

6.1.3 Head Loss of Laminar and Turbulent Flows in Pipe

On the experimental pipe segment, as the fluid in the horizontal straight pipeline keeps stable, according to the energy equation, the friction head loss equals to the pressure head difference between two sections. Namely

$$h_f = \frac{p_1 - p_2}{\gamma}$$

Change the velocity and measure v and the corresponding h_f under laminar flow and turbulent flow one by one, and the experimental results are shown in Figure 6-2.

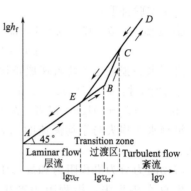

Figure 6-2　The Head Loss of Laminar and Turbulent Flows

图 6-2　层流、紊流的水头损失

The results show: Whether it is laminar flow or turbulent flow, the experimental points are all concentrated on the straight line with different slope. The equation is

$$\lg h_f = \lg k + m \lg v$$

where　$\lg k$ —intercept of a straight line;

　　　　m —slope of a straight line, and $m = \tan\theta$ (θ is the angle between the straight line and horizontal).

Many experiments prove:

① For laminar flow:

$\theta = 45°$, $m = 1$, namely

$$\lg h_f = \lg k_1 + \lg v \quad \text{or} \quad h_f = k_1 v$$

So it is indicated the friction head loss is in direct ratio with mean velocity of flow.

② For turbulent flow:

$\theta_1 > 45°$, $m = 1.75 \sim 2$, namely

$$\lg h_f = \lg k_2 + m \lg v \quad \text{or} \quad h_f = k_2 v^m$$

So it is indicated the friction head loss is in direct ratio with the 1.75 to 2 power of the mean velocity of flow.

Flow resistance includes friction and local resistance, thereinto, friction resistance produces friction head loss, while local resistance produces minor head loss. As shown in Figure 6-3, total head loss h_w is the summation of friction head loss h_f and minor head loss h_j.

$$h_w = \Sigma h_f + \Sigma h_j \tag{6-4}$$

6.1.3　管中层流、紊流的水头损失

在实验管段上，当水平直管内的流体为稳定流时，根据能量方程，其沿程水头损失就等于两断面间的压力水头差，即

$$h_f = \frac{p_1 - p_2}{\gamma}$$

改变速度，逐次测量层流、紊流两种情况下的 v 与对应的 h_f 值，实验结果如图 6-2 所示。结果表明：无论是层流状态还是紊流状态，实验点都分别集中在不同斜率的直线上，方程式为

$$\lg h_f = \lg k + m \lg v$$

式中，$\lg k$ 为直线的截距；m 为直线的斜率，且 $m = \tan\theta$（θ 为直线与水平线的交角）。

大量实验证明：

① 层流时：$\theta = 45°$，$m = 1$，即

$$\lg h_f = \lg k_1 + \lg v \quad \text{或者} \quad h_f = k_1 v$$

沿程水头损失与平均流速成正比。

② 紊流时：$\theta > 45°$，$m = 1.75 \sim 2$，即

$$\lg h_f = \lg k_2 + m \lg v \quad \text{或者} \quad h_f = k_2 v^m$$

沿程水头损失与平均流速的 $1.75 \sim 2$ 次方成正比。

流动阻力分为沿程阻力和局部阻力，流动阻力产生总水头损失，沿程阻力产生沿程水头损失，局部阻力产生局部水头损失。如图 6-3 所示，总水头损失 h_w 为沿程水头损失 h_f 与局部水头损失 h_j 之和。

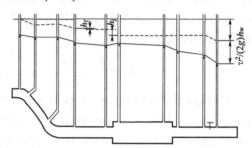

Figure 6-3　Flow Resistance and Head Loss
图 6-3　流动阻力及水头损失

$$h_w = \Sigma h_f + \Sigma h_j \tag{6-4}$$

6.2　Basic Equation of Uniform Flow

Friction head loss is a work to overcome friction resistance (tangential force). It is necessary to discuss and develop the relation between friction resistance and head loss, namely, basic equation of uniform flow. Therefore, firstly we must develop energy equation involving friction head loss and balance equation of forces including tangential force along the direction of stream.

There is only friction head loss in a uniform flow. As shown in Figure 6-4, in order to determine the friction head loss between section 1-1 and 2-2, the energy equation can be written as follows：

$$z_1 + \frac{p_1}{\gamma} + \frac{\alpha_1 v_1^2}{2g} = z_2 + \frac{p_2}{\gamma} + \frac{\alpha_2 v_2^2}{2g} + h_f$$

In a uniform flow,

$$\frac{\alpha_1 v_1^2}{2g} = \frac{\alpha_2 v_2^2}{2g}$$

So we can get

$$h_f = \left(z_1 + \frac{p_1}{\gamma}\right) - \left(z_2 + \frac{p_2}{\gamma}\right) \tag{6-5}$$

Stream of fluid between section 1-1 and 2-2 in a round pipe uniform flow is shown in Figure 6-4. Its length is l, area of the cross-section $A = A_1 = A_2$, and wetted perimeter is χ. Now let us analyze the conditions of force balance.

The flow is uniform because of the collective action of pressure force P_1 at section 1-1, pressure force P_2 at section 2-2, fluid weight G, and tangential force T on fluid surface. Write the balance equation of the force projections at the stream direction：

$$P_1 - P_2 + G\cos\alpha - T = 0$$

For $P_1 = p_1 A$, $P_2 = p_2 A$, and $\cos\alpha = \dfrac{z_1 - z_2}{l}$, given that the mean shear stress on contact face between liquid and solid wall is τ_0, from the equation above we obtain

$$p_1 A - p_2 A + \gamma A l \dfrac{z_1 - z_2}{l} - \tau_0 \chi l = 0$$

Dividing the equation above by γA yields

$$\dfrac{p_1}{\gamma} - \dfrac{p_2}{\gamma} + z_1 - z_2 = \dfrac{\tau_0 \chi}{\gamma A} l$$

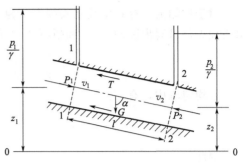

Figure 6-4　A Uniform Flow in a Round Pipe

图 6-4　圆管均匀流

From Equation (6-5), we know $\left(z_1 + \dfrac{p_1}{\gamma}\right) - \left(z_2 + \dfrac{p_2}{\gamma}\right) = h_f$.

Therefore

$$h_f = \dfrac{\tau_0 \chi}{\gamma A} l = \dfrac{\tau_0 l}{\gamma R} \tag{6-6}$$

or

$$\tau_0 = \gamma R \dfrac{h_f}{l} = \gamma R J \tag{6-7}$$

Equation (6-6) and Equation (6-7), which present the relation between friction head loss and shear stress, are the basic equations to study friction head loss, so they are called basic equations of uniform flow. For a gravitational uniform flow, according to the method above we can also get Equation (6-6) and Equation (6-7). So Equation (6-6) and Equation (6-7) are applicable to both pressure and gravitational flows.

For a round pipe, hydraulic radius $R = r_0/2$, where r_0 is radius of the round pipe. Given that a random cylindrical flow where the centre of a circle coincides with axis of pipe has a radius of r and the shear stress on surface of cylindrical flow is τ, we can obtain two equations as follows.

$$\tau_0 = \gamma \dfrac{r_0}{2} J \tag{6-8}$$

$$\tau = \gamma \dfrac{r}{2} J \tag{6-9}$$

Compared Equation (6-8) with Equation (6-9), we can get

$$\dfrac{\tau}{\tau_0} = \dfrac{r}{r_0} \tag{6-10}$$

Equation (6-10) shows at the cross-section in a round pipe uniform flow, the shear stress has a linear distribution, the maximum τ_0 exists on the wall of pipe and the shear stress at the pipe axis is equal to zero.

6.2　均匀流基本方程式

　　沿程水头损失是克服沿程阻力，即切向力所做的功。因此有必要讨论并建立沿程阻力和水头损失的关系——均匀流基本方程式，为此首先要建立包括沿程水头损失的能量方程及包括切向力在内的沿水流方向力的平衡方程。

对于均匀流只存在沿程水头损失。为了确定均匀流在断面 1—1 和断面 2—2 之间的沿程水头损失，可写出断面 1—1 和断面 2—2 的能量方程式（图 6-4）。

$$z_1+\frac{p_1}{\gamma}+\frac{\alpha_1 v_1^2}{2g}=z_2+\frac{p_2}{\gamma}+\frac{\alpha_2 v_2^2}{2g}+h_\mathrm{f}$$

在均匀流中，有

$$\frac{\alpha_1 v_1^2}{2g}=\frac{\alpha_2 v_2^2}{2g}$$

因此

$$h_\mathrm{f}=\left(z_1+\frac{p_1}{\gamma}\right)-\left(z_2+\frac{p_2}{\gamma}\right) \tag{6-5}$$

取出自过流断面 1—1 至 2—2 的一段圆管均匀流动的液流（图 6-4），其长度为 l，过水断面面积 $A=A_1=A_2$，湿周为 χ，现分析其作用力的平衡条件。

断面 1—1 至 2—2 间的流段是在断面 1—1 上的流动压力 P_1、断面 2—2 上的流动压力 P_2、流段本身的重量 G 及流段表面切向力 T 的共同作用下保持均匀流动的。

写出在流动方向上诸力投影的平衡方程式：

$$P_1-P_2+G\cos\alpha-T=0$$

因 $P_1=p_1A$，$P_2=p_2A$，而且 $\cos\alpha=\dfrac{z_1-z_2}{l}$，并设液流与固体边壁接触面上的平均切应力为 τ_0，代入上式，得

$$p_1A-p_2A+\gamma Al\frac{z_1-z_2}{l}-\tau_0\chi l=0$$

以 γA 除全式，得

$$\frac{p_1}{\gamma}-\frac{p_2}{\gamma}+z_1-z_2=\frac{\tau_0\chi}{\gamma A}l$$

由式（6-5）知 $\left(z_1+\dfrac{p_1}{\gamma}\right)-\left(z_2+\dfrac{p_2}{\gamma}\right)=h_\mathrm{f}$。

于是

$$h_\mathrm{f}=\frac{\tau_0\chi}{\gamma A}l=\frac{\tau_0 l}{\gamma R} \tag{6-6}$$

或

$$\tau_0=\gamma R\frac{h_\mathrm{f}}{l}=\gamma RJ \tag{6-7}$$

式（6-6）及式（6-7）给出了沿程水头损失与切应力的关系，是研究沿程水头损失的基本公式，称为均匀流基本方程式。对于无压均匀流，按上述步骤，列出沿流动方向的力平衡方程式，同样可得与式（6-6）及式（6-7）相同的结果，所以该方程对有压流和无压流均适用。

一圆管的水力半径 $R=\dfrac{r_0}{2}$，其中 r_0 为圆管半径。由于在圆管中取任一半径为 r 的圆柱流，列出作用力平衡方程，圆柱流圆心与管轴重合，圆柱流表面切应力为 τ，亦可得出以下两式

$$\tau_0=\gamma\frac{r_0}{2}J \tag{6-8}$$

$$\tau = \gamma \frac{r}{2} J \tag{6-9}$$

比较式(6-8)和式(6-9)，可得

$$\frac{\tau}{\tau_0} = \frac{r}{r_0} \tag{6-10}$$

式(6-10)说明在圆管均匀流的过水断面上，切应力呈直线分布，管壁处切应力为最大值 τ_0，管轴处切应力为零。

6.3 Laminar Flow in a Round Pipe

Laminar flow in a round pipe is known as Hagen-Poiseuille flow.

To derive the applicable equation to calculate the friction head loss from basic equation of uniform flow, we must further investigate the relation between shear stress τ and mean velocity v. τ depends on type of fluid flow, so we first analyze laminar flow in a round pipe.

6.3.1 Velocity Distribution and Mean Velocity

If the flow is laminar, shear stress between fluid layers can be calculated from Newton's friction law, namely

$$\tau = \mu \frac{du}{dy}$$

As the pressure uniform flow in a round pipe is an axial symmetry flow, to calculate expediently, cylindrical coordinates r, x (Figure 6-5) are used and the flow is a two-dimensional flow.

In the formula above, $y = r_0 - r$, so $dy = -dr$. Thus

$$\frac{du}{dy} = -\frac{du}{dr}$$

$$\tau = -\mu \frac{du}{dr} \tag{6-11}$$

Again the shear stress at the radius of r in a round uniform flow can be calculated by Equation (6-9), namely

$$\tau = \frac{1}{2} r \gamma J$$

From the two formulas above, we get

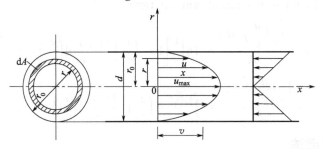

Figure 6-5 Velocity and Shear Stress Distribution of Laminar Flow in a Round Pipe

图 6-5 圆管层流的速度分布与切应力分布

$$\tau = -\mu \frac{du}{dr} = \frac{1}{2}r\gamma J$$

$$du = -\frac{\gamma}{2} \times \frac{J}{\mu}r\,dr$$

where J is a constant in each element flow in a uniform flow. Integrating the above formula yields velocity of a random element flow (velocity at a point).

$$u = -\frac{\gamma J}{4\mu}r^2 + C$$

$u = 0$ if $r = r_0$. Integral constant C is determined by this boundary condition, namely

$$C = \frac{\gamma J}{4\mu}r_0^2$$

Therefore

$$u = \frac{\gamma J}{4\mu}(r_0^2 - r^2) \tag{6-12}$$

Equation (6-12) shows the distribution of velocity at cross-section in a round pipe uniform flow is a paraboloid of revolution.

From Equation (6-12), the maximum velocity in flow obtained at axis of pipe, is

$$u_{max} = \frac{\gamma J}{4\mu}r_0^2 \tag{6-13}$$

For $Q = \int_A u\,dA = vA$, given that the area of annular section with a breadth of dr is dA, we can get the mean velocity in a round laminar flow:

$$v = \frac{Q}{A} = \frac{\int_A u\,dA}{A} = \frac{1}{\pi r_0^2}\int_0^{r_0}\frac{\gamma J}{4\mu}(r_0^2 - r^2)\,2\pi r\,dr = \frac{\gamma J}{8\mu}r_0^2 \tag{6-14}$$

Compared Equation (6-13) with Equation (6-14), we obtain

$$v = \frac{1}{2}u_{max} \tag{6-15}$$

Namely, in a round laminar flow the mean velocity is a half of the maximum velocity. Dividing Equation (6-12) by Equation (6-14), we get a dimensionless formula:

$$\frac{u}{v} = 2\left[1 - \left(\frac{r}{r_0}\right)^2\right] \tag{6-16}$$

Equation (6-16) can be used to calculate the kinetic-energy correction factor α and momentum correction factor β in a round laminar flow.

$$\alpha = \frac{\int_A \left(\frac{u}{v}\right)^3 dA}{A} = 16\int_0^1 \left[1 - \left(\frac{r}{r_0}\right)^2\right]^3 \frac{r}{r_0}\,d\left(\frac{r}{r_0}\right) = 2$$

$$\beta = \frac{\int_A \left(\frac{u}{v}\right)^2 dA}{A} = 8\int_0^1 \left[1 - \left(\frac{r}{r_0}\right)^2\right]^2 \frac{r}{r_0}\,d\left(\frac{r}{r_0}\right) = 1.33$$

6.3 圆管中的层流

圆管中的层流运动也称为哈根-泊肃叶（Hagen-Poiseuille）流动。要由均匀流基本方程推出沿程水头损失的计算公式，必须进一步研究切应力 τ 与平均流速 v 的关系，而 τ 的变化与流体的流态有关，因此我们需要先对圆管中的层流运动进行分析。

6.3.1 圆管层流的速度分布及平均流速

流体在层流运动时，液层间的切应力可由牛顿内摩擦定律求出，该式为

$$\tau = \mu \frac{du}{dy}$$

圆管中有压均匀流是轴对称流，为了计算方便，采用圆柱坐标 r、x（图 6-5），此时为二元流。

上式中的 $y = r_0 - r$，故 $dy = -dr$，因此

$$\frac{du}{dy} = -\frac{du}{dr}$$

$$\tau = -\mu \frac{du}{dr} \tag{6-11}$$

圆管均匀流在半径 r 处的切应力可用均匀流方程 [式(6-9)] 表示，即

$$\tau = \frac{1}{2} r \gamma J$$

由以上两式得

$$\tau = -\mu \frac{du}{dr} = \frac{1}{2} r \gamma J$$

于是

$$du = -\frac{\gamma}{2} \times \frac{J}{\mu} r dr$$

注意 J 对均匀流中各元流都是不变的，积分上式得任一元流的流速（点流速）：

$$u = -\frac{\gamma J}{4\mu} r^2 + C$$

当 $r = r_0$ 时，$u = 0$，用该边界条件确定积分常数 C：

$$C = \frac{\gamma J}{4\mu} r_0^2$$

所以

$$u = \frac{\gamma J}{4\mu} (r_0^2 - r^2) \tag{6-12}$$

式(6-12) 说明圆管层流运动断面上的流速分布是一个旋转抛物面。

流动中的最大速度在管轴上，由式(6-12) 得

$$u_{\max} = \frac{\gamma J}{4\mu} r_0^2 \tag{6-13}$$

因为流量 $Q = \int_A u dA = vA$，选取宽为 dr 的环形断面，令该微元的面积为 dA，可得圆管层流运动的平均流速：

$$v = \frac{Q}{A} = \frac{\int_A u dA}{A} = \frac{1}{\pi r_0^2} \int_0^{r_0} \frac{\gamma J}{4\mu} (r_0^2 - r^2) 2\pi r dr = \frac{\gamma J}{8\mu} r_0^2 \tag{6-14}$$

比较式(6-13) 和式(6-14)，得

$$v = \frac{1}{2} u_{\max} \tag{6-15}$$

即圆管层流的平均流速为最大流速的一半。

由式(6-12)比式(6-14)得无量纲关系式：

$$\frac{u}{v} = 2\left[1 - \left(\frac{r}{r_0}\right)^2\right] \tag{6-16}$$

利用式(6-16)可计算圆管层流的动能修正系数 α 和动量修正系数 β：

$$\alpha = \frac{\int_A \left(\frac{u}{v}\right)^3 dA}{A} = 16\int_0^1 \left[1 - \left(\frac{r}{r_0}\right)^2\right]^3 \frac{r}{r_0} d\left(\frac{r}{r_0}\right) = 2$$

$$\beta = \frac{\int_A \left(\frac{u}{v}\right)^2 dA}{A} = 8\int_0^1 \left[1 - \left(\frac{r}{r_0}\right)^2\right]^2 \frac{r}{r_0} d\left(\frac{r}{r_0}\right) = 1.33$$

6.3.2 Distribution of Shear Stress

According to Newton's law of internal friction, for flow in round pipe we can obtain

$$\tau = -\mu \frac{du_x}{dy} = -\mu \frac{du_x}{dr} = \frac{\Delta p r}{2l} \tag{6-17}$$

Equation (6-17) can be written as the following equation:

$$\tau = \tau(r) \tag{6-18}$$

when $r = R$, the shear stress on the pipe wall is

$$\tau_0 = \frac{\Delta p R}{2l} \tag{6-19}$$

So

$$\frac{\tau}{\tau_0} = \frac{r}{R}$$

This shows the shear stress at the cross section of laminar flow is in direct ratio with the radius. The distribution rule is shown in Figure 6-5 and the distribution is called k-shaped.

6.3.2 切应力分布

根据牛顿内摩擦定律，对于圆管中的流动，有

$$\tau = -\mu \frac{du_x}{dy} = -\mu \frac{du_x}{dr} = \frac{\Delta p r}{2l} \tag{6-17}$$

式(6-17)可写成如下形式：

$$\tau = \tau(r) \tag{6-18}$$

当 $r = R$ 时，可得管壁处的切应力为

$$\tau_0 = \frac{\Delta p R}{2l} \tag{6-19}$$

则

$$\frac{\tau}{\tau_0} = \frac{r}{R}$$

说明在层流的过流断面上，切应力与半径成正比，分布规律如图 6-5 所示，称为切应力的 k 形分布。

6.3.3 Friction Loss for a Laminar Flow

According to Bernoulli's equation we know that the head loss of flow in a constant diameter pipe is equal to the difference of pressure head on the two ends of pipe.

$$h_f = \frac{\Delta p}{\rho g} = \frac{8\nu l Q}{\pi g R^4} = \frac{128\nu l Q}{\pi g d^4} \qquad (6\text{-}20)$$

or

$$h_f = \frac{8\nu l v}{g R^2} = \frac{32\nu l v}{g d^2} \qquad (6\text{-}21)$$

Equation (6-21) shows that the energy loss of laminar flow is in proportion to the first power of v.

Dimensional analysis shows the friction loss can be expressed by Equation (6-22).

$$h_f = \lambda \frac{l}{d} \times \frac{v^2}{2g} \qquad (6\text{-}22)$$

Equation (6-22) is known as the Darcy-Weisbach formula, and in this formula, λ is the friction resistance coefficient.

According to the Darcy-Weisbach formula, the friction head loss in a round pipe is expressed by this formula for laminar or turbulent flow. Comparing with Equation (6-21), the friction resistance coefficient of laminar flow is

$$\lambda = \frac{64\nu}{vd} = \frac{64}{Re} \qquad (6\text{-}23)$$

so

$$h_f = \lambda \frac{l}{d} \times \frac{v^2}{2g} = \frac{64}{Re} \times \frac{l}{d} \times \frac{v^2}{2g}$$

For friction head loss, Equation (6-22) is the most common and basic formula.

6.3.3 层流的沿程损失

根据伯努利方程可知，等径管路的水头损失就是管路两端的压强水头之差，即

$$h_f = \frac{\Delta p}{\rho g} = \frac{8\nu l Q}{\pi g R^4} = \frac{128\nu l Q}{\pi g d^4} \qquad (6\text{-}20)$$

或

$$h_f = \frac{8\nu l v}{g R^2} = \frac{32\nu l v}{g d^2} \qquad (6\text{-}21)$$

式(6-21)说明层流的能量损失与 v 的一次方成比例。

量纲分析表明，沿程阻力损失可用式(6-22)表示。

$$h_f = \lambda \frac{l}{d} \times \frac{v^2}{2g} \qquad (6\text{-}22)$$

式(6-22)即为达西-魏斯巴赫公式，式中 λ 为沿程阻力系数。

根据达西-魏斯巴赫公式，不论层流还是紊流，圆管中的沿程水头损失都可用该式表示。与式(6-21)相比，可得层流的沿程阻力系数：

$$\lambda = \frac{64\nu}{vd} = \frac{64}{Re} \qquad (6\text{-}23)$$

于是有

$$h_f = \lambda \frac{l}{d} \times \frac{v^2}{2g} = \frac{64}{Re} \times \frac{l}{d} \times \frac{v^2}{2g}$$

式(6-22)所表示的沿程水头损失是最常用、最基本的一种形式。

6.4 Turbulent Flow in a Round Pipe

When Reynolds number exceeds critical value, the fluid flow in pipe will turn into turbulent flow.

6.4.1 The Occurrence of Turbulent Flow

The velocity difference between two flow layers is the basic reason of causing the instability. If the fluid is disturbed gently, the unstable laminar flow will turn into turbulent flow.

As shown in Figure 6-6 (a), there are many layers of linear flow with different velocities. If the interface is disturbed gently, as shown in Figure 6-6 (b), the velocity at one point decreases but pressure increases. At the same time the pressure at another point decreases. Thus, due to the pressure difference, the fluid particles on the interface will flow from one point to another point, and then the disturbance on the interface will be intensified. Gradually, it expands to turbulent flow and eddies will be formed as shown in Figure 6-6 (c).

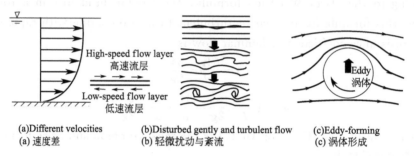

Figure 6-6   The Occurrence of Turbulent Flow
图 6-6   紊流的产生

6.4 圆管中的紊流

当雷诺数超过临界值时，管内流体流动变成紊流。

6.4.1 紊流的产生

两层流体间有速度差别，是造成不稳定的根本原因，不稳定的层流受到轻微扰动即可转化为紊流。图 6-6(a) 所示为速度不同的直线流动。如分界面受轻微扰动，见图 6-6(b)，某一点处速度降低而压强增大，同时另一点处压强则下降，界面处的流体质点由于压差将由某一点向另一点流动，加剧界面的扰动，逐渐向紊流发展并形成涡体，见图 6-6(c)。

6.4.2 Fluctuation of Turbulent Flow

After laminar flow is destroyed, many vortices with different magnitudes and directions are formed in a turbulent flow. These vortices are the reason to result in the velocity fluctuation. As shown in Figure 6-7, for an adequate long period, it is found that the fluctuation always varies around a mean value.

① According to the variational curve of velocity at a point in Figure 6-7, substitute the instantaneous value with the mean value of time at the period of time T, and we obtain

$$\bar{u} = \frac{1}{T}\int_0^T u_t \mathrm{d}t \tag{6-24}$$

where $\bar{u}$ — time mean velocity.

② At different periods, the difference between the factual velocity u and time mean velocity $\bar{u}$ is expressed by u' and is called fluctuating velocity. Then

$$u(t) = \bar{u} + u'(t) \tag{6-25}$$

Note: $u'(t)$ can be positive or negative, but its mean value of time $\overline{u'}(t) = 0$. Other velocity components and pressure all can be expressed by the mean value of time in turbulent flow.

6.4.2 紊流的脉动

层流破坏以后，在紊流中形成许多大大小小方向不同的旋涡，这些旋涡是造成速度脉动的原因。

在足够长的时间内，人们发现脉动总是围绕着某一平均值而变化，如图 6-7 所示。

① 根据图 6-7 所示的一点上的速度变化曲线，用 T 时间段内的时间平均值代替瞬时值，则

$$\bar{u} = \frac{1}{T} \int_0^T u_t \, \mathrm{d}t \tag{6-24}$$

$\bar{u}$ 被称为一点上的时均速度。

② 不同时刻实际流速 u 与时均速度的差值用 u' 表示，称为脉动流速，则有

$$u(t) = \bar{u} + u'(t) \tag{6-25}$$

注: $u'(t)$ 值可正可负，但其时均值 $\overline{u'}(t) = 0$。紊流中的其他流速分量和压强也都可类似地以时均值表示。

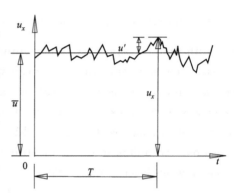

Figure 6-7 The Velocity Fluctuation at a Point in a Turbulent Flow

图 6-7 紊流中一点上的速度脉动

6.4.3 Mixing Length Theory

Prandtl founded the mixing length theory to explain the influence of fluctuation on the time mean flow.

Considering round pipe flow and plane flow, as shown in Figure 6-8, a coordinate system is chosen. For the round pipe flow, direction of y is the reverse direction of coordinate r, and the maximum value of y is the radius R of the round pipe. The mean value of velocity on plane or in a round pipe can be expressed by $u = \bar{u}(y)$. Assume there are two flow layers a and b in the time mean flow. The mean value of velocity on layer a is $\bar{u}$ and the mean value of velocity on layer b is $\bar{u} + l\dfrac{\mathrm{d}\bar{u}}{\mathrm{d}y}$.

Assume that at a certain time there is a fluid element on layer a where the mean value of velocity is $\bar{u}$. For some occasional factors it jumps up along y with fluctuating velocity u'_y through area of element $\mathrm{d}A$. The mass flux is $\rho u'_y \mathrm{d}A$. Prandtl considered that $\bar{u}$ is constant before the fluid element reaching a new posi-

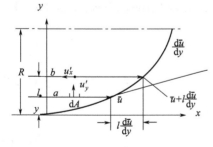

Figure 6-8 Mixing Length

图 6-8 混合长度

tion. When it goes through the distance l and arrives at the layer b where the time mean velocity is $\bar{u}+l\dfrac{d\bar{u}}{dy}$, it mixes with the fluid of layer b immediately. And then it has the time mean velocity $\bar{u}+l\dfrac{d\bar{u}}{dy}$ on layer b.

Because the original momentum $\rho u'_y \bar{u} dA$ in the x direction of the fluid element is smaller than the momentum $\rho u'_y (\bar{u}+l\dfrac{d\bar{u}}{dy}) dA$ at the time it reaches the layer b, so when it mixes with fluid on layer b, the momentum in the x direction of total fluid on layer b will decrease. That is to say, the time mean velocity in the x direction will decrease. Thus there appears an instantaneous fluctuation $-u'_x$ on layer b (minus expresses that the direction is reverse with the x axis). Because of the new fluctuating velocity $-u'_x$, a new fluctuating momentum change $\rho u'_y (-u'_x - 0) dA$ occurs to the fluid element mixing on layer b in the x direction. According to the momentum theorem, the momentum change will cause the shear force between layer a and b. So

$$F = -\rho u'_y u'_x dA$$

The shear stress between layer a and layer b is

$$\tau = -\rho u'_x u'_y \qquad (6\text{-}26)$$

This is the Reynolds shear stress caused by fluctuation.

When $u'_y > 0$, the element fluctuates from layer a to layer b, fluctuating velocity on layer b $u'_x < 0$; when $u'_y < 0$, the element fluctuates from layer b to layer a, fluctuating velocity on layer a $u'_x > 0$. The symbols of u'_x and u'_y are always opposite, so $u'_x u'_y < 0$. Therefore $\tau = -\rho u'_x u'_y > 0$. In the time mean flow, Reynolds shear stress will not disappear and its mean value of time is

$$\bar{\tau} = \dfrac{1}{T}\int_0^T \tau dt = -\rho \dfrac{1}{T}\int_0^T u'_x u'_y dt = -\rho \overline{u'_x u'_y} \qquad (6\text{-}27)$$

This shows that though Reynolds shear stress caused by the fluctuation is a fluctuating variable, it has mean value of time and it has some influences on flow.

Prandtl considered that u'_x and u'_y are in proportion to each other, namely

$$u'_y = -k_1 u'_x \qquad (6\text{-}28)$$

u'_x is in proportion to the difference of time mean velocities between layer a and b, that is $l\dfrac{d\bar{u}}{dy}$.

$$u'_x = k_2 l \dfrac{d\bar{u}}{dy} \qquad (6\text{-}29)$$

From Equation (6-28) and Equation (6-29), we obtain

$$u'_y = -k_1 k_2 l \dfrac{d\bar{u}}{dy} \qquad (6\text{-}30)$$

Substituting Equation (6-29) and Equation (6-30) into Equation (6-27) yields

$$\bar{\tau} = \rho k_1 k_2^2 l^2 \left(\dfrac{d\bar{u}}{dy}\right)^2$$

Let $k_1 k_2^2 = k^2$, and then mean value of fluctuating shear stress is

$$\bar{\tau} = \rho (kl)^2 \left(\frac{d\bar{u}}{dy}\right)^2 = \rho L^2 \left(\frac{d\bar{u}}{dy}\right)^2 \tag{6-31}$$

where L is called mixing length, and $L = kl$.

6.4.3 混合长度理论

普朗特（Prandtl）创立了混合长度理论，合理地解释了脉动对时均流动的影响。为了兼顾圆管与平面流动这两种情况，如图 6-8 所示选取一平面坐标系。对于圆管来说 y 轴方向就是 r 坐标的反方向，y 可取的最大值就是圆管半径 R，平面或圆管断面上的时均速度分布都可以用 $u = \bar{u}(y)$ 表示。

设在时均流动中有 a、b 两层流体，a 层的时均速度为 $\bar{u}$，b 层的时均速度为 $\bar{u} + l\dfrac{d\bar{u}}{dy}$。设想在某一瞬时，在时均速度为 $\bar{u}$ 的 a 层上有一个流体微团，由于某种偶然因素，经过微元面积 dA 以 u'_y 的脉动速度沿 y 轴正向跳动，其质量流量为 $\rho u'_y dA$。普朗特认为在流体微团到达新的位置之前，它原来具有的 $\bar{u}$ 一直不变，当它经过 l 距离到达时均速度为 $\bar{u} + l\dfrac{d\bar{u}}{dy}$ 的 b 层以后，立即与 b 层流体混合在一起，从而具有 b 层的时均速度 $\bar{u} + l\dfrac{d\bar{u}}{dy}$。

这个流体微团原来所具有的 x 方向的动量 $\rho u'_y \bar{u} dA$ 小于它到 b 层后所具有的 x 方向的动量 $\rho u'_y (\bar{u} + l\dfrac{d\bar{u}}{dy}) dA$，因而它与 b 层流体混合后，必然使整个 b 层流体在 x 方向上的动量有所降低，也就是使其 x 方向上的时均速度有所降低，这样在 b 层上就出现了一个瞬时的速度脉动 $-u'_x$（"$-$"号表示它的方向与 x 轴相反）。新产生的脉动速度 $-u'_x$，使混合到 b 层的这个流体微团在 x 方向上产生了一个新的脉动性的动量变化 $\rho u'_y (-u'_x - 0) dA$。按照动量定理，这个动量变化必然引起 a、b 两层之间的切向作用力 F，所以

$$F = -\rho u'_y u'_x dA$$

a、b 两层之间的切应力为

$$\tau = -\rho u'_x u'_y \tag{6-26}$$

这就是由脉动引起的雷诺切应力。当 $u'_y > 0$，微团由 a 层向 b 层脉动，b 层的 $u'_x < 0$；当 $u'_y < 0$，微团由 b 层向 a 层脉动，a 层的 $u'_x > 0$。u'_x 与 u'_y 永远反号，因而 $u'_x u'_y < 0$，所以 $\tau = -\rho u'_x u'_y > 0$，在时均化的过程中，雷诺切应力并不消失，它的时均值为

$$\bar{\tau} = \frac{1}{T}\int_0^T \tau dt = -\rho \frac{1}{T}\int_0^T u'_x u'_y dt = -\rho \overline{u'_x u'_y} \tag{6-27}$$

这说明，脉动产生的雷诺切应力虽然是个脉动量，但它存在时均值，对流动产生确定的影响。

普朗特认为 u'_x 与 u'_y 互成比例，即

$$u'_y = -k_1 u'_x \tag{6-28}$$

u'_x 与 a、b 两层的时均速度之差 $l\dfrac{d\bar{u}}{dy}$ 也成正比，即

$$u'_x = k_2 l \frac{d\bar{u}}{dy} \tag{6-29}$$

由式(6-28)、式(6-29) 可得

$$u'_y = -k_1 k_2 l \frac{d\overline{u}}{dy} \tag{6-30}$$

将式(6-29)、式(6-30) 代入式(6-27) 中，得

$$\overline{\tau} = \rho k_1 k_2^2 l^2 \left(\frac{d\overline{u}}{dy}\right)^2$$

令 $k_1 k_2^2 = k^2$，则脉动切应力的时均值为

$$\overline{\tau} = \rho (kl)^2 \left(\frac{d\overline{u}}{dy}\right)^2 = \rho L^2 \left(\frac{d\overline{u}}{dy}\right)^2 \tag{6-31}$$

式中，$L = kl$ 称为混合长度。

6.4.4 Distributions of Shear Stress and Velocity of Turbulent Flow in Pipe

The turbulent flow structures in viscous sub-layer, hydraulic smooth round pipe and hydraulic rough round pipe are shown in Figure 6-9.

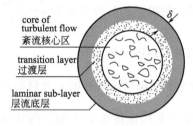

Figure 6-9 Turbulent Flow Structures
图 6-9 紊流结构

Viscous force predominates near the pipe wall where the mixing is confined to form laminar layer, which is called laminar sub-layer. The layers out of laminar sub-layer are transition layer and out of transition layer is the core of turbulent flow.

When the turbulent flow is fully developed, there are two states near the wall: If Reynolds number is smaller and turbulent flow sub-layer covers the coarse protuberance on pipe wall, the roughness does nothing to the turbulent flow, as shown in Figure 6-10 (a). That is called hydraulic smooth pipe. With the increase of Reynolds number, turbulent flow sub-layer turns thin. When the coarse protuberance is higher than the turbulent flow sub-layer, the coarse protuberance can cause turbulent flow quickly. The higher coarse protuberance is, the bigger the resistance is, as shown in Figure 6-10 (b). That is called hydraulic rough pipe.

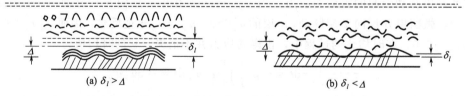

Figure 6-10 Hydraulic Smooth Pipe (a) and Hydraulic Rough Pipe (b)
图 6-10 水力光滑管 (a) 和水力粗糙管 (b)

The depth of laminar sub-layer δ can be determined approximately by the following formula:

$$\delta \approx 30 \frac{d}{Re\sqrt{\lambda}} \tag{6-32}$$

where δ——depth of laminar sub-layer;
d——inner diameter of pipe;

Re — Reynolds number;

λ — hydraulic friction coefficient (different with laminar flow).

(1) Shear stress distribution For the time mean turbulent flow, each point in fluid has only an axial time mean velocity u_x in pipe. Shear stress on the pipe wall is

$$\tau_0 = \frac{(p_1 - p_2)R}{2l} = \frac{\Delta p}{l} \times \frac{R}{2} \tag{6-33}$$

where R — radius of pipe;

Δp — the pressure difference of two sections between which the axial distance is l.

If we take out a flow pipe of which the radius is r ($r < R$) between the two sections, then in a same way we can obtain the shear stress on the surface of flow pipe.

$$\tau = \frac{\Delta p}{l} \times \frac{r}{2} \tag{6-34}$$

From Equation (6-33) and Equation (6-34), we can get

$$\tau = \tau_0 \frac{r}{R} \tag{6-35}$$

This is the distribution regulation of shear stress at cross-section of pipe.

(2) Velocity distribution In the viscous sub-layer, $\tau = \mu \dfrac{du_x}{dy}$. Rearranging the formula, $du_x = \dfrac{\tau}{\mu} dy$. Because laminar sub-layer is very thin, τ can be expressed by shear stress τ_0 approximately. So after integrating

$$u_x = \frac{\tau_0}{\mu} y \tag{6-36}$$

As shown in Figure 6-11, the velocity has a linear distribution in laminar sub-layer. This is the approximate result of parabola distribution in laminar sub-layer.

In the core of turbulent flow, $\tau = \rho L^2 \left(\dfrac{du_x}{dy}\right)^2$. From Equation (6-35), we obtain

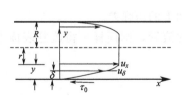

Figure 6-11 Velocity Distribution in Turbulent Flow

图 6-11 紊流的速度分布

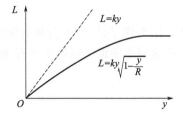

Figure 6-12 Mixing Length Distribution

图 6-12 混合长度分布

$$\tau = \tau_0 \frac{r}{R} = \tau_0 \left(1 - \frac{y}{R}\right) \tag{6-37}$$

According to Karman experiment, the distribution of mixing length is shown in Figure 6-12, and the function formula between L and y can be expressed approximately as follows:

$$L = ky \sqrt{1 - \frac{y}{R}} \tag{6-38}$$

Near the wall, $y \ll R$, thus

$$L = ky \tag{6-39}$$

in which k is an empirical constant and $k = 0.4$.

Substitute Equation (6-37) and Equation (6-38) into fluctuating shear stress expression

$$\tau = \rho L^2 \left(\frac{du_x}{dy}\right)^2$$

Simplifying and integrating, we obtain

$$u_x = \sqrt{\frac{\tau_0}{\rho}} \frac{1}{k} \ln y + C \tag{6-40}$$

This formula shows that velocity in the core of turbulent flow u_x is in logarithm relation with y.

6.4.4 管中紊流的切应力分布和速度分布

黏性底层、水力光滑与水力粗糙圆管中的紊流结构如图 6-9 所示。

在靠近管壁处，黏性力占优势，此处混合受限制，形成层流层，称为层流底层。层流底层外面紧接的是过渡层；过渡层外面紧接的是紊流核心区。

完全紊流时，近壁处存在两种状态：雷诺数较小时，近壁处层流底层完全掩盖住管壁上的粗糙突起，此时粗糙度对紊流不起作用，如图 6-10(a) 所示，称为水力光滑管；随雷诺数增大，层流底层变薄，当粗糙突起高出层流底层时，粗糙突起造成紊动加剧，粗糙突起突出越高，阻力越大，如图 6-10(b) 所示，称为水力粗糙管。

层流底层厚度 δ 近似地可用下式确定：

$$\delta \approx 30 \frac{d}{Re\sqrt{\lambda}} \tag{6-32}$$

式中，δ 为层流底层厚度；d 为管内径；Re 为雷诺数；λ 为水力摩擦系数（与层流时不同）。

(1) 切应力分布 对时均化的紊流来说，流体每一点在管中只有一个轴向时均速度 u_x，管壁上的切应力为

$$\tau_0 = \frac{(p_1 - p_2)R}{2l} = \frac{\Delta p}{l} \times \frac{R}{2} \tag{6-33}$$

式中，R 为管半径；Δp 为轴向距离为 l 的两断面上的压强差。

如果在这两个断面之间取出半径为 r ($r < R$) 的流管，则同样可得流管表面上的切应力为

$$\tau = \frac{\Delta p}{l} \times \frac{r}{2} \tag{6-34}$$

由式(6-33) 及式(6-34) 两式可得

$$\tau = \tau_0 \frac{r}{R} \tag{6-35}$$

这就是过流断面上切应力的分布规律。

(2) 速度分布 在黏性底层中 $\tau = \mu \frac{du_x}{dy}$，即 $du_x = \frac{\tau}{\mu} dy$。

因为层流底层很薄，τ 可近似用壁面上的切应力 τ_0 表示。于是积分可得

$$u_x = \frac{\tau_0}{\mu} y \tag{6-36}$$

如图 6-11 所示，在层流底层中速度的分布呈直线规律，这是层流速度抛物线规律在层流底层中的近似结果。

在紊流核心中，$\tau = \rho L^2 \left(\dfrac{\mathrm{d}u_x}{\mathrm{d}y}\right)^2$，由式(6-35) 得

$$\tau = \tau_0 \frac{r}{R} = \tau_0 \left(1 - \frac{y}{R}\right) \tag{6-37}$$

根据卡门（Karman）实验，混合长度的分布规律如图 6-12 所示，L 与 y 的函数关系可近似表示为

$$L = ky \sqrt{1 - \frac{y}{R}} \tag{6-38}$$

当 $y \ll R$，即在壁面附近时

$$L = ky \tag{6-39}$$

其中，$k = 0.4$，为经验常数。

将式(6-37)、式(6-38) 代入脉动切应力的表达式

$$\tau = \rho L^2 \left(\frac{\mathrm{d}u_x}{\mathrm{d}y}\right)^2$$

化简、积分得

$$u_x = \sqrt{\frac{\tau_0}{\rho}} \times \frac{1}{k} \ln y + C \tag{6-40}$$

说明紊流核心中速度 u_x 和 y 成对数关系。

6.5 Friction Resistance in Pipeline

Friction resistance is the reason that causes the friction head (pressure, energy) loss. Darcy's formula is used to calculate the friction loss, but the coefficient of friction resistance $\lambda = f\left(Re, \dfrac{\Delta}{d}\right)$ needs further discussion.

6.5 管路中的沿程阻力

沿程阻力是造成沿程水头（或压强、能量）损失的原因。计算沿程损失的公式是达西公式，但式中沿程阻力系数 $\lambda = f\left(Re, \dfrac{\Delta}{d}\right)$ 的规律有待深入探讨。

6.5.1 Nikuradse Experiment

Nikuradse daubed the sifted sands with the same diameter on the inner wall of pipeline with different diameters to make a manual rough pipeline and carried out an experiment. Reynolds number of experiment $Re = 500 \sim 10^6$, and relative roughness was $\Delta/d = 1/1014 \sim 1/30$. The experimental curve is shown in Figure 6-13.

According to Figure 6-13, there exist five different zones due to the relations between λ, Re and Δ/d as follows.

① Laminar flow zone. As $Re < 2300$, all experimental points concentrate on a straight

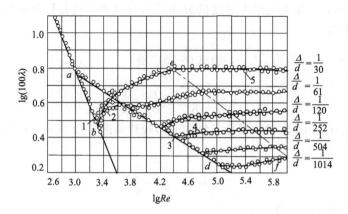

Figure 6-13 Nikuradse Experimental Curve
图 6-13 尼古拉兹实验曲线

line ab, which shows that λ has nothing to do with relative roughness Δ/d. The relation between λ and Re fits equation $\lambda = 64/Re$ which is consistent completely with the laminar theory formula of round pipe.

② Transition zone. Transition zone is a zone in which laminar flow turns into turbulent flow. In this zone λ has nothing to do with Δ/d, as shown in zone 2 of Figure 6-13.

③ Turbulent smooth pipe zone. As $Re > 2300$, though flow keeps turbulent, experimental points with different roughness all concentrate on line cd, which shows that roughness has nothing to do with λ, but only has relation with Reynolds number Re. The depth of laminar sub-layer is bigger than roughness of pipe, $\delta > \Delta$.

④ Transition zone of turbulent flow. With the increase of Reynolds number experimental points leave from line cd according to the roughness of different points respectively, and go into the transition zone of turbulent flow, as shown in zone 4 of Figure 6-13.

⑤ Complete rough pipe zone or the resistance square zone. The zone of Figure 6-13, in which the experimental curve parallels with x axis, $\Delta > \delta$, and friction resistance is in direct ratio with the square of velocity, is called complete rough pipe zone or the resistance square zone. It can be seen from Figure 6-13 that in this zone λ has nothing to do with Re but only has relation with roughness Δ/d.

The limit ranges of five resistance zones and their λ calculating formulas are listed in Table 6-1.

Table 6-1 The Ranges of Five Resistance Zones and Their λ Calculating Formulas
表 6-1 五个阻力区的范围及其 λ 计算公式

resistance zone (阻力区)	range (范围)	the theoretical λ or half-empirical formula of λ (λ 的理论或半经验公式)	empirical formula of λ (λ 的经验公式)
laminar flow zone (层流区)	$Re < 2300$	$\lambda = 64/Re$	$\lambda = 75/Re$
transition zone (过渡区)	$2300 < Re < 3000$	—	$\lambda = 0.0025 Re^{\frac{1}{3}}$ $Re < 10^5$

续表

resistance zone (阻力区)	range (范围)	the theoretical λ or half-empirical formula of λ (λ 的理论或半经验公式)	empirical formula of λ (λ 的经验公式)
turbulent smooth pipe zone (紊流光滑管区)	$3000 < Re < 22.2\left(\dfrac{d}{\Delta}\right)^{\frac{8}{7}}$	$\dfrac{1}{\sqrt{\lambda}} = 2\lg(Re\sqrt{\lambda}) - 0.8$	$\lambda = \dfrac{0.3164}{Re^{0.25}}$ $10^5 < Re < 3 \times 10^6$ $\lambda = 0.0032 + \dfrac{0.221}{Re^{0.237}}$
transition zone of turbulent flow (紊流过渡区)	$22.2\left(\dfrac{d}{\Delta}\right)^{\frac{8}{7}} < Re < 597\left(\dfrac{d}{\Delta}\right)^{\frac{9}{8}}$	$\dfrac{1}{\sqrt{\lambda}} = -2\lg\left(\dfrac{\Delta}{3.7d} + \dfrac{2.51}{Re\sqrt{\lambda}}\right)$	$\lambda = 0.11\left(\dfrac{\Delta}{d} + \dfrac{68}{Re}\right)^{0.25}$
complete rough pipe zone / the resistance square zone (粗糙管区/阻力平方区)	$Re > 597\left(\dfrac{d}{\Delta}\right)^{\frac{9}{8}}$	$\lambda = \dfrac{1}{\left[2\lg\left(3.7\dfrac{d}{\Delta}\right)\right]^2}$	$\lambda = 0.11\left(\dfrac{\Delta}{d}\right)^{0.25}$

The half-empirical formulas of Table 6-1 are obtained by coordinating experimental data on the basis of the mixing length theory and velocity distribution. They have more accuracy but the structures are complicated. The accuracy of empirical formula in the last column is a little worse but it is easy to calculate. Sometimes we use empirical formula to calculate the first approximation and then substitute it into the right side of half-empirical formula of smooth pipe or transition zone of turbulent flow, and then calculate the second approximation. If we substitute it into the right side again, we can calculate the third approximation from the left side. Iterating two or three times we can obtain the accurate value which makes the left side is equal to the right side.

6.5.1 尼古拉兹实验

尼古拉兹将不同管径的管道内壁均匀地粘涂上经过筛分具有相同粒径的砂粒，以制成人工粗糙管道进行实验研究，实验范围雷诺数 $Re = 500 \sim 10^6$，相对粗糙度 $\Delta/d = 1/1014 \sim 1/30$，实验曲线如图 6-13 所示。

由图 6-13 可以看出，λ 和 Re 及 $\dfrac{\Delta}{d}$ 的关系可分为五个不同的区，其变化规律如下。

① 层流区。当 $Re < 2300$，所有的实验点都聚集在一条直线 ab 上，说明 λ 与相对粗糙度 $\dfrac{\Delta}{d}$ 无关，而 λ 与 Re 的关系符合方程 $\lambda = \dfrac{64}{Re}$，这与圆管层流理论公式完全一致。

② 过渡区。该区是层流转变为紊流的过渡区，此时 λ 与 $\dfrac{\Delta}{d}$ 无关，如图 6-13 中的区域 2 所示。

③ 紊流光滑管区。当 $Re > 2300$，流动虽已处于紊流状态，但不同粗糙度的实验点都聚集在 cd 线上，说明粗糙度对 λ 仍没有影响，只与雷诺数 Re 有关。层流底层厚度大于管子粗糙度，$\delta > \Delta$。

④ 紊流过渡区。随着雷诺数的增大，实验点根据不同点的粗糙度分别从 cd 线上离开，进入紊流过渡区，如图 6-13 中区域 4 所示。

⑤ 粗糙管区或阻力平方区。图中实验曲线与横轴平行的区域，$\Delta > \delta$，沿程阻力与速度

的平方成正比，称为粗糙管区或阻力平方区，从图 6-13 中可以看出在此区域 λ 与 Re 无关，而仅与粗糙度 $\dfrac{\Delta}{d}$ 有关。

五个阻力区的界限范围及其 λ 计算公式汇总列于表 6-1 中。表中半经验公式是建立在混合长度理论及速度分布的基础上并配合实验数据而得到的，它们的准确性较高，但是结构较复杂。最末一栏的经验公式准确性稍差，但公式简单，便于计算。有时也可以先用经验公式求第一次近似值，然后将其代入光滑管或紊流过渡区的半经验公式右端，从其左端求出第二次近似值。如果再将第二次近似值代入右端，则从左端又可求出第三次近似值。迭代两三次即可得左、右基本相等的准确值。

6.5.2 Moody Diagram

Depending on forementioned formulas and many experimental data, Moody made the diagram in Figure 6-14 on friction loss coefficient λ with Reynolds number Re and relative roughness Δ/d for industry pipeline. According to this figure, we can expediently calculate the value of loss coefficient λ and judge the resistance zone.

6.5.2 穆迪图

穆迪（Moody）依据大量实验资料，并借助于前述各公式对工业用管道制作了关于沿程阻力系数 λ 与雷诺数 Re 和相对粗糙度 $\dfrac{\Delta}{d}$ 的图 6-14。根据此图可以很方便地求得沿程阻力系数 λ 的值，并可以判断所在的阻力区。

【**Sample Problem 6-2**】 Water (20℃) flows in the welded steel pipe with a diameter of 50cm. If the energy loss for unit length of pipe is 0.006, try to calculate the discharge and

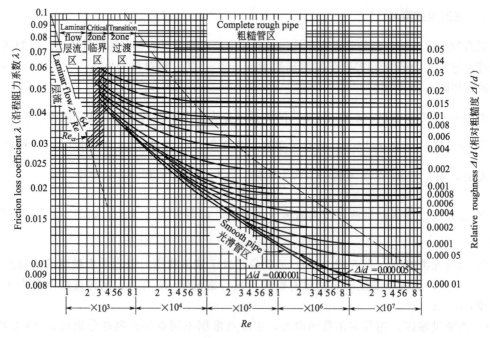

Figure 6-14　Moody Diagram
图 6-14　穆迪图

depth of viscous sub-layer in pipe ($\Delta/d = 0.046/500 = 0.00009$).

Solution: From Formula (6-22), we obtain

$$\frac{h_f}{l} = 0.006 = \lambda \frac{1}{d} \times \frac{v^2}{2g} = \lambda \frac{1}{0.5} \times \frac{v^2}{2g}$$

From the formula above, $v = 0.243/\sqrt{\lambda}$.

Assume $\lambda = 0.030$, and then $v = 1.4$ m/s. For $\nu = 1.007 \times 10^{-6}$ m²/s,

$$Re = \frac{dv}{\nu} = \frac{0.5 \times 1.4}{1.007 \times 10^{-6}} \approx 7 \times 10^5$$

Check from Moody Diagram $\left(Re = 7 \times 10^5, \frac{\Delta}{d} = 0.00009\right)$, $\lambda = 0.0136$, and then

$$v = \frac{0.243}{\sqrt{0.0136}} = 2.08 \text{m/s}$$

$$Re = \frac{0.5 \times 2.08}{1.007 \times 10^{-6}} \approx 10^6$$

Check it again, $\lambda = 0.0131$, $v = \frac{0.243}{\sqrt{\lambda}} = 2.12$ m/s, $Re \approx 10^6$.

So $v = 2.12$ m/s is the velocity.

Then

$$Q = Av = \frac{\pi (0.5)^2}{4} \times 2.12 = 0.416 \text{m}^3/\text{s}$$

Again from Equation (6-32),

$$\delta = \frac{30\nu}{v\sqrt{\lambda}} = \frac{30 \times (1.007 \times 10^{-6})}{2.12 \times \sqrt{0.0131}} = 125 \times 10^{-6} \text{m} = 0.125 \text{mm}$$

【例题 6-2】 20℃的水在管径为 50cm 的焊接钢管内流动,若单位管长的能量损失为 0.006,试计算管中流量、黏性底层厚度$\left(\frac{\Delta}{d} = 0.046/500 = 0.00009\right)$。

解: 由式(6-22) 得

$$\frac{h_f}{l} = 0.006 = \lambda \frac{1}{d} \times \frac{v^2}{2g} = \lambda \frac{1}{0.5} \times \frac{v^2}{2g}$$

由上式得到 $v = 0.243/\sqrt{\lambda}$。

设 $\lambda = 0.030$,则 $v = 1.4$ m/s, $\nu = 1.007 \times 10^{-6}$ m²/s,

$$Re = \frac{dv}{\nu} = \frac{0.5 \times 1.4}{1.007 \times 10^{-6}} \approx 7 \times 10^5$$

由穆迪图查得$\left(Re = 7 \times 10^5, \frac{\Delta}{d} = 0.00009\right)$, $\lambda = 0.0136$,则

$$v = \frac{0.243}{\sqrt{0.0136}} = 2.08 \text{m/s}$$

$$Re = \frac{0.5 \times 2.08}{1.007 \times 10^{-6}} \approx 10^6$$

再查穆迪图, $\lambda = 0.0131$, $v = \frac{0.243}{\sqrt{\lambda}} = 2.12$ m/s, $Re \approx 10^6$。

可见 $v = 2.12$ m/s 为所求的速度,则

$$Q = Av = \frac{\pi(0.5)^2}{4} \times 2.12 = 0.416 \text{m}^3/\text{s}$$

再由式(6-32)，

$$\delta = \frac{30\nu}{v\sqrt{\lambda}} = \frac{30 \times (1.007 \times 10^{-6})}{2.12 \times \sqrt{0.0131}} = 125 \times 10^{-6} \text{m} = 0.125 \text{mm}$$

6.6 Minor Resistance in Pipeline

Minor resistance will occur and then cause minor head loss where the flow sections change quickly or fluid flow direction changes. Though there are many sorts of pipe fittings in pipeline, the causes of head loss include:

① redistribution of velocity of flow in fluids;
② viscous force doing work in swirling;
③ momentum change caused by the mix of liquid particles.

Because of the quick variation of border, the turbulent degree of fluid flow is enforced. So the minor loss is in direct ratio with the square of mean velocity of flow normally, and it can be expressed by

$$h_j = \xi \frac{v^2}{2g} \tag{6-41}$$

where h_j—minor loss;
ξ—coefficient of minor loss.

6.6 管路中的局部阻力

在液流断面急剧变化以及液流方向转变的地方，易产生局部阻力，引起局部水头损失，虽然管路上安装的各种管件多种多样，但产生局部水头损失的原因不外乎以下几种。

① 液流中流速的重新分布。
② 在旋涡中黏性力做功。
③ 液体质点掺混引起的动量变化。

边界的急剧变化，加强了流体流动的紊动程度，故局部损失一般和平均流速的平方成正比，可表达为

$$h_j = \xi \frac{v^2}{2g} \tag{6-41}$$

式中，h_j 为局部损失；ξ 为局部损失系数。

6.6.1 Minor Head Loss on Sudden Expansion of Pipe Section

We can use theory analysis to confirm the minor head loss. The most representative instance is the sudden expansion of pipe section.

As shown in Figure 6-15, the fluid occurs to swirling via the position where the section enlarges suddenly, and after length l main fluid enlarges to all section. Section 1-1 and 2-2 can be considered as gradually-varying flow section. Because the distance between section 1-1 and 2-2 is very short, its friction loss can be ignored. Then apply Bernoulli's equation:

$$h_j = \left(z_1 + \frac{p_1}{\rho g} + \frac{\alpha_1 v_1^2}{2g}\right) - \left(z_2 + \frac{p_2}{\rho g} + \frac{\alpha_2 v_2^2}{2g}\right)$$
$$= \left(z_1 + \frac{p_1}{\rho g}\right) - \left(z_2 + \frac{p_2}{\rho g}\right) + \left(\frac{\alpha_1 v_1^2}{2g} - \frac{\alpha_2 v_2^2}{2g}\right) \tag{6-42}$$

We can apply momentum equation to fluid in the control surface $AB22$. Firstly we analyze components of outer forces along flow direction of fluid in the control surface $AB22$.

① Whole pressure acting on section 1-1 is $p_1 A_1$, where p_1 is pressure on axes;

② Whole pressure acting on section 2-2 is $p_2 A_2$, where p_2 is pressure on axes;

③ Acting force of ringy area pipe wall of AB ($A_2 - A_1$) is the reverse force of the swirling acting on the ringy area. Experiment shows that the distributions of pressures on ringy area obey the static pressure rule $F = p_1 (A_2 - A_1)$.

④ Component of fluid gravity in control surface via flow direction is
$$G\cos\theta = \rho g A_2 l (z_1 - z_2)/l = \rho g A_2 (z_1 - z_2)$$

⑤ Compared with the above forces, the friction resistance of fluid between section AB and 2-2 can be ignored.

Momentum equation in control surface $AB22$ is
$$\rho Q (v_2 - v_1) = p_1 A_1 - p_2 A_2 + p_1 (A_2 - A_1) + \rho g A_2 (z_1 - z_2)$$

Substituting $Q = v_2 A_2$ into the equation above and divide both sides by $\rho g A_2$, we obtain
$$\frac{v_2}{g}(v_2 - v_1) = \left(z_1 + \frac{p_1}{\rho g}\right) - \left(z_2 + \frac{p_2}{\rho g}\right) \tag{6-43}$$

Substituting Equation (6-43) into Equation (6-42) yields $h_j = \dfrac{(v_1 - v_2)^2}{2g}$.

This equation is minor loss formula for round pipe sudden expansion. According to continuity equation $\rho A_1 v_1 = \rho A_2 v_2$, the equation above can be written as
$$h_j = \left(\frac{A_2}{A_1} - 1\right)^2 \frac{v_2^2}{2g} = \xi_2 \frac{v_2^2}{2g}$$

or
$$h_j = \left(1 - \frac{A_1}{A_2}\right)^2 \frac{v_1^2}{2g} = \xi_1 \frac{v_1^2}{2g} \tag{6-44}$$

Thus,
$$\xi_1 = \left(1 - \frac{A_1}{A_2}\right)^2, \quad \xi_2 = \left(1 - \frac{A_2}{A_1}\right)^2 \tag{6-45}$$

It should be explained that most of the minor resistance loss coefficients can be confirmed by experiment.

6.6.1 圆管中水流突然扩大的局部水头损失

借助理论分析来确定局部水头损失时，最有代表性的是管路突然扩大的情况。

如图 6-15 所示，由于流体经突然扩大处产生旋涡，经过长度 l 后主流扩大到整个断面，断面 1—1 及断面 2—2 可认为是渐变流断面，又因 1—1 与 2—2 断面间的距离较短，其沿程损失可忽略不计，则应用伯努利方程得

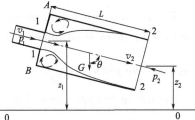

Figure 6-15 Sudden Expansion of Pipe Section

图 6-15 管道截面突然扩大

$$h_j = \left(z_1 + \frac{p_1}{\rho g} + \frac{\alpha_1 v_1^2}{2g}\right) - \left(z_2 + \frac{p_2}{\rho g} + \frac{\alpha_2 v_2^2}{2g}\right) \tag{6-42}$$

$$= \left(z_1 + \frac{p_1}{\rho g}\right) - \left(z_2 + \frac{p_2}{\rho g}\right) + \left(\frac{\alpha_1 v_1^2}{2g} - \frac{\alpha_2 v_2^2}{2g}\right)$$

再对控制面 $AB22$ 内的流体运用动量方程。首先分析控制面 $AB22$ 内流体所受外力沿流动方向的分力，有：

① 作用在断面 1—1 上的总压力 $p_1 A_1$，其中 p_1 为轴线上的压强。

② 作用在断面 2—2 上的总压力 $p_2 A_2$，其中 p_2 为轴线上的压强。

③ AB 环形面积（$A_2 - A_1$）管壁对流体的作用力，即旋涡作用于环形面积上的反力，实验表明，环形面积上压强按静压强规律分布，即总压力 $F = p_1 (A_2 - A_1)$。

④ 控制面内流体重力沿流动方向的分力为

$$G\cos\theta = \rho g A_2 l (z_1 - z_2)/l = \rho g A_2 (z_1 - z_2)$$

⑤ 断面 AB 至 2—2 间流体所受管壁的摩擦阻力与上述诸力相比可忽略不计。

控制面 $AB22$ 动量方程为

$$\rho Q(v_2 - v_1) = p_1 A_1 - p_2 A_2 + p_1 (A_2 - A_1) + \rho g A_2 (z_1 - z_2)$$

代入 $Q = v_2 A_2$，并除以 $\rho g A_2$ 得

$$\frac{v_2}{g}(v_2 - v_1) = \left(z_1 + \frac{p_1}{\rho g}\right) - \left(z_2 + \frac{p_2}{\rho g}\right) \tag{6-43}$$

将式(6-43)代入式(6-42)得

$$h_j = \frac{(v_1 - v_2)^2}{2g}$$

此式即为圆管突然扩大的局部损失公式。根据连续性方程 $\rho A_1 v_1 = \rho A_2 v_2$，上式又可写成

$$h_j = \left(\frac{A_2}{A_1} - 1\right)^2 \frac{v_2^2}{2g} = \xi_2 \frac{v_2^2}{2g}$$

或

$$h_j = \left(1 - \frac{A_1}{A_2}\right)^2 \frac{v_1^2}{2g} = \xi_1 \frac{v_1^2}{2g} \tag{6-44}$$

从上式可知

$$\xi_1 = \left(1 - \frac{A_1}{A_2}\right)^2, \quad \xi_2 = \left(1 - \frac{A_2}{A_1}\right)^2 \tag{6-45}$$

应当说明的是，绝大多数局部阻力损失系数可由实验确定。

6.6.2 Minor Head Loss on Sudden Contraction of Pipe Section

The sudden contraction of pipe section is shown in Figure 6-16. Due to different swirl zone in big pipe or small pipe, if the mean velocity v_2 in the small pipe is used to measure head loss, we get $h_j = \xi \frac{v_2^2}{2g}$.

According to experimental results

$$\xi = 0.5\left(1 - \frac{A_2}{A_1}\right) \tag{6-46}$$

Equation (6-46) shows the minor head loss of sudden contraction of pipe is less than that of corresponding sudden expansion pipe.

6.6.2 断面突然收缩的水头损失

断面突然收缩的情形如图 6-16 所示。在大管和小管中都有不同的旋涡区，如用小管中的平均流速 v_2 来衡量水头损失，则

$$h_j = \xi \frac{v_2^2}{2g}$$

根据实验结果

$$\xi = 0.5\left(1 - \frac{A_2}{A_1}\right) \tag{6-46}$$

通过计算比较可知，突然收缩的水头损失比相应的突然放大的要小。

6.6.3 Minor Head Loss at Pipe Entrance or Exit

As seen in Figure 6-16, pipe entrance could be thought as a special case where the section 1-1 is very large, namely, A_2/A_1 is approximate to zero, so from Equation (6-46) we can get $\xi = 0.5$.

When the entrance is bell-mouthed, according to stream condition at entrance, we can choose the range: $\xi = 0.05 \sim 0.25$.

The minor loss coefficient at pipe exit $\xi = 1.0$.

6.6.3 管路进出口水头损失

管路进口可以看作图 6-16 中 1—1 断面非常大的一种特殊情况，即 A_2/A_1 近似于零，则从式(6-46)可得 $\xi = 0.5$。如为喇叭形进口，则视进口水流情况，可选择如下范围的系数值：$\xi = 0.05 \sim 0.25$。

管道出口局部损失系数 $\xi = 1.0$。

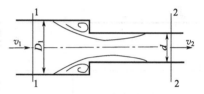

Figure 6-16 Minor Head Loss of Sudden Contraction Pipe

图 6-16 突然缩小管局部水头损失

6.6.4 Minor Head Loss of Gradual Expansions and Contractions, Elbows, Bends and Other Pipe Fittings

The formulas for calculating minor loss coefficients and the values of gradual expansions and contractions, elbows, bends and other pipe fittings are listed in Table 6-2, 6-3, 6-4, 6-5 and 6-6, respectively.

6.6.4 渐扩、渐缩、弯折管及管路配件局部水头损失

渐扩、渐缩、弯管、折管及管路配件的局部水头损失系数的计算公式及数值分别列在表 6-2、表 6-3、表 6-4、表 6-5 及表 6-6 中。

Table 6-2 Minor Loss Coefficients of Gradual Expansion Pipe

表 6-2 断面逐渐扩大管损失系数

D/d	Gradual expansion pipe 断面逐渐扩大管 $h_j = \zeta \dfrac{(v_1-v_2)^2}{2g}$											
	2°	4°	6°	8°	10°	15°	20°	25°	30°	35°	40°	45°
1.1	0.01	0.01	0.01	0.02	0.03	0.05	0.10	0.13	0.16	0.18	0.19	0.20
1.2	0.02	0.02	0.02	0.03	0.06	0.09	0.16	0.21	0.25	0.29	0.31	0.33
1.4	0.02	0.03	0.03	0.04	0.06	0.12	0.23	0.30	0.36	0.41	0.44	0.47

续表

D/d	Gradual expansion pipe 断面逐渐扩大管 $h_j = \zeta \dfrac{(v_1-v_2)^2}{2g}$											
	2°	4°	6°	8°	10°	15°	20°	25°	30°	35°	40°	45°
1.6	0.03	0.03	0.04	0.05	0.07	0.14	0.26	0.35	0.42	0.47	0.51	0.54
1.8	0.03	0.04	0.04	0.05	0.07	0.15	0.28	0.37	0.44	0.50	0.54	0.58
2.0	0.03	0.04	0.04	0.05	0.07	0.16	0.29	0.38	0.45	0.52	0.56	0.60
2.5	0.03	0.04	0.04	0.05	0.08	0.16	0.30	0.39	0.48	0.54	0.58	0.62
3.0	0.03	0.04	0.04	0.05	0.08	0.16	0.31	0.40	0.48	0.55	0.59	0.63

Table 6-3 Minor Loss Coefficients of Gradual Contraction Pipe
表 6-3 断面逐渐缩小管损失系数

Gradual contraction pipe 断面逐渐缩小管
$h_m = \zeta \dfrac{v_2^2}{2g}$

D_2/D_1	0.0	0.1	0.2	0.3	0.4	0.5
ζ	0.50	0.45	0.42	0.39	0.36	0.33
D_2/D_1	0.6	0.7	0.8	0.9	1.0	
ζ	0.28	0.22	0.15	0.06	0.00	

Table 6-4 Minor Loss Coefficients of Elbows
表 6-4 弯管损失系数

$$\zeta = \left[0.131 + 0.163\left(\dfrac{d}{R}\right)^{0.5}\right]\left(\dfrac{\theta}{90°}\right)^{0.5}$$
Round pipe(圆管)

Slowly-bend pipe (缓弯管)			ζ				
	90°	d/R	0.2	0.4	0.6	0.8	1.0
		ζ	0.132	0.138	0.158	0.206	0.294
		d/R	1.2	1.4	1.6	1.8	2.0
		ζ	0.440	0.660	0.976	1.406	1.975
	90°	b/R	0.2	0.4	0.6	0.8	1.0
		ζ	0.12	0.14	0.18	0.25	0.40
		b/R	1.2	1.4	1.6	2.0	
		ζ	0.64	1.02	1.55	3.23	
Arbitrary angle 任意角度		θ	15°	30°	45°	60°	120°	150°	180°
		$k = \left(\dfrac{\theta}{90°}\right)^{0.5}$	0.41	0.57	0.71	0.82	1.16	1.29	1.41

Table 6-5 Minor Loss Coefficients of Bends
表 6-5 折管损失系数

Bend（折管）

$$\zeta = 0.946\sin^2\left(\dfrac{\theta}{2}\right) + 2.05\sin^2\left(\dfrac{\theta}{2}\right)$$
（圆管）

Round pipe（圆管）	θ	20°	45°	60°	90°	120°
	ζ	0.045	0.183	0.365	0.99	1.86
Square pipe（方管）	θ	15°	30°	45°	60°	90°
	ζ	0.025	0.11	0.260	0.490	1.2

Table 6-6　Minor Loss Coefficients of Pipe Fittings
表 6-6　管路配件局部损失系数

其他管路配件局部损失 $h_j = \zeta \dfrac{v^2}{2g}$

Types 种类	Schema 图式	ζ	Types 种类	Schema 图式	ζ
Globe valve 截止阀		Wide-open 全开　4.3~6.1	Straight tee 等径三通		0.1
Butterfly valve 蝶阀		Wide-open 全开　0.1~0.3			1.5
Valve 阀门		Wide-open 全开　0.12			1.5
Valveless water filter net 无阀滤水网		2~3			3.0
Bottom valve with mesh 有网底阀		3.5~10 (d=50~600mm)			2.0

Exercises 习题

6-1　In a refinery, oil ($s=0.85$, $\nu=1.8\times10^{-5}\,\mathrm{m^2/s}$) flows through a pipe with a diameter of 100mm at 0.50L/s. Is the flow laminar or turbulent?

6-1　在一家炼油厂，石油以 0.50L/s 的流量流经直径为 100mm 的管道，石油的相对密度为 $s=0.85$，运动黏度为 $\nu=1.8\times10^{-5}\,\mathrm{m^2/s}$，试判别该流动是层流还是紊流。

6-2　What is the hydraulic radius of a rectangular water pipe with a width of 1.2m and a length of 1.6m?

6-2　宽和长分别是 1.2m 和 1.6m 的矩形水管的水力半径是多少？

6-3　What is the percentage difference between the hydraulic radius of a round pipe with a diameter of 500mm and a square pipe with a side length of 500mm?

6-3　直径为 500mm 的圆管和边长为 500mm 的正方形管道的水力半径相差多少？用百分比表示。

6-4　There are two pipes with the same cross-sectional area. One is circular and the other one is square. Which has the larger hydraulic radius, and by what percentage?

6-4　有两个管道，一个圆形，一个正方形，并且有相同的横截面积，哪一个管道的水力半径比较大？大多少？用百分比表示。

Figure for Exercise 6-4
习题 6-4 图

6-5　In the laminar flow of a fluid through a pipe of circular cross-section the velocity profile is exactly a true parabola. The volume of the paraboloid represents the rate of discharge. Prove that for this case the ratio of the mean velocity to the maximum velocity is 0.5.

6-5　经过圆形截面管道的液体层流速度分布图是一条抛物线，旋转抛物体的体积代表流量。求证在这种情况下，平均速度与最大速度的比值为 0.5。

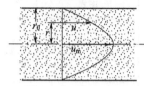

Figure for Exercise 6-5
习题 6-5 图

6-6　Velocities in a 200-mm-diameter circular conduit, measured at radii of 0mm, 36mm, 65mm, and 87mm, were 7.0m/s, 6.8m/s, 6.1m/s and 5.0m/s, respectively. Find approximate values (graphically) of the volume flow rate and the mean velocity. And determine the ratio of the mean velocity to the maximum velocity.

6-6　直径为 200mm 的圆形管道，以半径为 0mm、36mm、65mm 和 87mm 测量，测得速度分别为 7.0m/s、6.8m/s、6.1m/s 和 5.0m/s，用图解法求体积流量和平均速度的近似值，同时求出平均速度和最大速度的比值。

6-7　In laminar flow through a circular pipe the velocity profile is a parabola, the equation of which is $u = u_m [1-(r/r_0)^2]$, where u is the velocity at any radius r, u_m is the maximum velocity in the center of the pipe where $r = 0$, and r_0 is the radius to the wall of the pipe. Find α.

6-7　经过圆形截面管道的液体层流速度分布图是一条抛物线，其表达式是 $u = u_m [1-(r/r_0)^2]$，其中 u 指半径为 r 时的速度，u_m 是当 $r = 0$ 时管道中心点的最大速度，r_0 是从管道中心到管壁的半径，求 α。

6-8　A 150m long pipe ($\lambda = 0.025$) with a diameter of 300mm connects one reservoir with another, both ends of the pipe being under water. The intake is square-edged. The difference between the water surface levels of the two reservoirs is 36m. Find: ① the flow rate; ② the pressure in the pipe at a point 100m from the intake, where the elevation is 39m lower than the surface of the water in the upper reservoir.

6-8　一根长 150m、直径为 300mm 的水管（$\lambda = 0.025$）连通两个蓄水池，水管的两端都在水面下，水管的入口处是直角边缘，两个蓄水池的水平面高度相差 36m，求：① 流量；② 距离入口 100m 且比高位的蓄水池的水平面低 39m 处的压强。

6-9　The diameter of a horizontal pipe with a butterfly valve is 100mm. In order to test the coefficient ξ of the minor head loss, as shown in Figure, a U-shaped pipe is set up. The flow rate of water in pipe $Q = 15.0$L/s and the mercury height is 50mm in the U-shaped

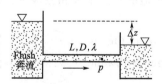

Figure for Exercise 6-8
习题 6-8 图

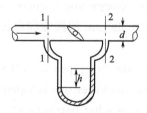

Figure for Exercise 6-9
习题 6-9 图

pipe. If friction head losses are neglected, try to calculate the coefficient ξ of minor head loss of the butterfly valve.

6-9 水平管直径 $d=100$mm，管中装一蝶阀，为测定蝶阀的局部水头损失系数 ξ，在蝶阀前后设有 U 形水银压差计，如图所示，当管中流量 $Q=15.0$L/s 时，水银柱高差 $h=50$mm，设沿程水头损失可以不计，求蝶阀在该开度时的局部水头损失系数 。

Chapter 7 Pressure Flow
第 7 章 有压管流

7.1 Single Pipeline

Long pipe or short pipe is not completely geometry concept of long and short, but a concept on resistance calculation. In pipe calculation many physical variables are concerned and many problems need to be solved. But the basic problems can be treated as three types:

① With l, d and Q known, calculate h_f;
② With l, d and h_f known, calculate Q;
③ With l, h_f and Q known, calculate d.

7.1.1 Short Pipe

Short pipe is one of the most familiar pipelines in mechanical engineering, especially oil pipes in mechanical equipment and water pipes in workshop. Usually their minor resistances can't be ignored, so we should consider both the friction resistance loss and minor resistance loss in pipe calculation.

The pipeline, in which both friction loss and minor loss are in a certain proportion, is called short pipe.

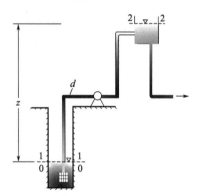

Figure 7-1 Pipeline of Water Pump
图 7-1 水泵管路

【Sample Problem 7-1】 Pipeline of water pump is shown in Figure 7-1, given that diameter of cast-iron pipe $d = 150$mm and length $l = 180$m. There is a filter-net ($\xi = 6$), an open globe valve, three elbows of which ratios of pipe radius to curvature radius are $r/R = 0.5$ in pipeline. The altitude $h = 100$m, discharge $Q = 225 \text{m}^3/\text{h}$, and temperature of water is 20℃. Try to calculate the output power of the water pump.

Solution: Firstly it is necessary to judge the flow type to confirm the friction resistance coefficient λ. The kinematic viscosity of water at 20℃ $\nu = 1.007 \times 10^{-6}$ m^2/s, so

$$Re = \frac{v d}{\nu} = \frac{4Q}{\pi d \nu} = \frac{4 \times 225 \times 10^6}{3600\pi \times 0.15 \times 1.007} = 5.27 \times 10^5$$

Cast-iron pipe:

$$\Delta = 0.25\text{mm}, \quad \frac{\Delta}{d} = 0.00166$$

$$\frac{d}{\Delta} = 600$$

$$22.2\left(\frac{d}{\Delta}\right)^{\frac{8}{7}} = 22.2 \times 600^{\frac{8}{7}} = 33200 < Re$$

So the flow is turbulent flow of non-smooth pipe.

$$597 \times \left(\frac{d}{\Delta}\right)^{\frac{9}{8}} = 7.97 \times 10^5 > Re$$

We can know that the flow type is transition zone of turbulent flow.

Use empirical formula to calculate the approximation of λ.

$$\lambda = 0.11\left(\frac{\Delta}{d} + \frac{68}{Re}\right)^{0.25} = 0.0227$$

Substitute this value into the right side of half-empirical formula and then calculate the second approximation of λ from the left side. So

$$\frac{1}{\sqrt{\lambda}} = -2\lg\left(\frac{\Delta}{3.7d} + \frac{2.51}{Re\sqrt{\lambda}}\right) = 6.212$$

We can obtain $\lambda = 0.02559$. It is near to the first approximation, so we use this value as a correct value.

From minor resistance coefficient table and given data in this sample problem, we know that entrance $\zeta_1 = 0.5$, elbow $\zeta_2 = 0.29443$, globe valve $\zeta_3 = 3.97$, filer-net $\zeta_4 = 6$, and exit $\zeta_5 = 1$,

$$v = \frac{4Q}{\pi d^2} = \frac{4 \times 225}{60 \times 60 \times \pi \times 0.15^2} = 3.539\text{m/s}$$

$$h_w = \Sigma h_f + \Sigma h_j = \lambda \frac{L}{d} \times \frac{v^2}{2g} + (\zeta_1 + 3\zeta_2 + \zeta_3 + \zeta_4 + \zeta_5)\frac{v^2}{2g}$$

$$= \lambda \frac{L}{d} \times \frac{v^2}{2g} + (\zeta_1 + 3\zeta_2 + \zeta_3 + \zeta_4 + \zeta_5)\frac{v^2}{2g}$$

$$= 0.02559 \times \frac{180}{0.15} \times \frac{3.539^2}{2g} + (0.5 + 3 \times 0.29443 + 3.97 + 6 + 1) \times \frac{3.539^2}{2g}$$

$$= 19.62 + 7.89 = 27.5\text{m}$$

Water pump lift is

$$H = z + h_w = 100 + 27.5 = 127.5\text{m}$$

So the output power of the water pump is

$$P = \rho g Q H = 9800 \times \frac{225}{3600} \times 127.5 = 78.09\text{kW}$$

7.1 简单管道

长管和短管并不完全是几何长短的概念,而是一个阻力计算上的概念。管路计算中涉及的物理量很多,需要解决的问题也很多,不过问题的基本类型不外乎以下三种:或是已知 l、d、Q 求 h_f,或是已知 l、d、h_f 求 Q,或是已知 l、h_f、Q 求 d。

7.1.1 短管

短管是机械工程中最常见的一种管路，尤其是机械设备上的油管、车间中的水管等，它们的局部阻力往往不能忽略，因此在计算中需要同时考虑沿程阻力损失和局部阻力损失。水头损失中沿程损失、局部损失各占一定比例，这种管路称为短管。

【例题 7-1】 水泵管路如图 7-1 所示，铸铁管直径 $d=150\text{mm}$，长度 $l=180\text{m}$，管路上装有滤水网（$\xi=6$）一个，全开截止阀一个，管半径与曲率半径之比 $r/R=0.5$ 的弯头三个，高程 $h=100\text{m}$，流量 $Q=225\text{m}^3/\text{h}$，水温为 20℃。试求水泵输出功率。

解： 首先需要判断流动状态以便确定沿程阻力系数 λ，20℃时，水的运动黏度 $\nu=1.007\times10^{-6}\text{m}^2/\text{s}$，于是

$$Re=\frac{vd}{\nu}=\frac{4Q}{\pi d\nu}=\frac{4\times225\times10^6}{3600\pi\times0.15\times1.007}=5.27\times10^5$$

对于铸铁管：

$$\Delta=0.25\text{mm},\quad \frac{\Delta}{d}=0.00166$$

$$\frac{d}{\Delta}=600$$

$$22.2\left(\frac{d}{\Delta}\right)^{\frac{8}{7}}=22.2\times600^{\frac{8}{7}}=33200<Re$$

所以管路中为非光滑管紊流。

$$597\times\left(\frac{d}{\Delta}\right)^{\frac{9}{8}}=7.97\times10^5>Re$$

可知流动状态为紊流过渡区。

先用经验公式求 λ 的近似值。

$$\lambda=0.11\left(\frac{\Delta}{d}+\frac{68}{Re}\right)^{0.25}=0.0227$$

将此值代入半经验公式的右端，从其左端求 λ 的第二次近似值，于是

$$\frac{1}{\sqrt{\lambda}}=-2\lg\left(\frac{\Delta}{3.7d}+\frac{2.51}{Re\sqrt{\lambda}}\right)=6.212$$

解出 $\lambda=0.02559$，与第一次近似值相差不多，即以此值为准。

从局部阻力系数表及题中给出数据可知：入口 $\zeta_1=0.5$，弯头 $\zeta_2=0.29443$，截止阀 $\zeta_3=3.97$，滤水网 $\zeta_4=6$，出口 $\zeta_5=1$，$v=\dfrac{4Q}{\pi d^2}=\dfrac{4\times225}{60\times60\times\pi\times0.15^2}=3.539\text{m/s}$。

$$h_w=\sum h_f+\sum h_j=\lambda\frac{L}{d}\frac{v^2}{2g}+(\zeta_1+3\zeta_2+\zeta_3+\zeta_4+\zeta_5)\frac{v^2}{2g}$$

$$=\lambda\frac{L}{d}\times\frac{v^2}{2g}+(\zeta_1+3\zeta_2+\zeta_3+\zeta_4+\zeta_5)\frac{v^2}{2g}$$

$$=0.02559\times\frac{180}{0.15}\times\frac{3.539^2}{2g}+(0.5+3\times0.29443+3.97+6+1)\times\frac{3.539^2}{2g}$$

$$=19.62+7.89=27.5\text{m}$$

水泵扬程

$$H=z+h_w=100+27.5=127.5\text{m}$$

水泵输出功率

$$P = \rho g Q H = 9800 \times \frac{225}{3600} \times 127.5 = 78.09 \text{kW}$$

7.1.2 Long Pipe

The pipeline, in which friction head loss is the most, and minor head loss and velocity head loss can be ignored, is called long pipe. As shown in Figure 7-2, applying energy equation between section 1 and 2, we can obtain

$$z_1 + \frac{p_1}{\gamma} + \frac{\alpha_1 v_1^2}{2g} = z_2 + \frac{p_2}{\gamma} + \frac{\alpha_2 v_2^2}{2g} + h_w$$

For
$$z_1 - z_2 = H$$
$$\frac{p_1}{\gamma} = \frac{p_2}{\gamma} = 0$$
$$\frac{\alpha_1 v_1^2}{2g} = \frac{\alpha_2 v_2^2}{2g} = 0$$
$$h_w = h_f$$

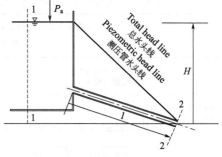

Figure 7-2 Head Line of Long Pipe
图 7-2 长管的水头线

Substituting into the equation above, we can obtain

$$H = h_f = \lambda \frac{l}{d} \times \frac{v^2}{2g}$$

Substituting $v = \frac{4Q}{\pi d^2}$ into the equation above yields $H = \frac{8\lambda}{g \pi^2 d^5} l Q^2$.

Let
$$A = \frac{8\lambda}{g \pi^2 d^5} \tag{7-1}$$

where A— specific resistance.

Thus
$$H = A l Q^2 \tag{7-2}$$

Specific resistance A denotes the requisite water head for unit flux flowing through unit pipeline length, and its magnitude depends on λ and d.

Specific resistance A can be obtained using the method to calculate λ introduced in previous sections. For old steel pipe and old cast-iron pipe, Шевелев equation can be used to calculate specific resistance A.

For resistance square zone ($v \geqslant 1.2 \text{m/s}$)

$$A = \frac{0.001736}{d^{5.3}}$$

For transition zone of turbulent flow ($v < 1.2 \text{m/s}$)

$$A' = 0.852 \left(1 + \frac{0.867}{v}\right)^{0.3} \left(\frac{0.001736}{d^{5.3}}\right) = kA$$

(7-3)

where k— a correction coefficient, and $k = 0.852 \left(1 + \frac{0.867}{v}\right)^{0.3}$.

At the temperature of 10℃, the k values at different velocities are listed in Table 7-1, and the specific resistance A values of steel pipe and old cast-iron pipe calculated by Equation (7-3) are listed in Table 7-2 and Table 7-3, respectively.

Table 7-1 k Values of Specific Resistance A of Steel Pipe and Old Cast-iron Pipe

表 7-1 钢管及铸铁管 A 值的修正系数 k

v/(m/s)	0.20	0.25	0.30	0.35	0.40	0.45	0.50	0.55	0.60
k	1.41	1.33	1.28	1.24	1.20	1.175	1.15	1.13	1.115
v/(m/s)	0.65	0.70	0.75	0.80	0.85	0.90	1.0	1.1	≥1.2
k	1.10	1.085	1.07	1.06	1.05	1.04	1.03	1.015	1.00

Table 7-2 Specific Resistance A of Steel Pipe (s^2/m^6)

表 7-2 钢管的比阻 A 值   单位：s^2/m^6

Water gas pipe (水煤气管)			Medium-sized pipe (中等管径)		Big-diameter pipe (大管径)	
Nominal diameter (公称直径) D_g/mm	A (Q 以 m^3/s 计)	A (Q 以 L/s 计)	Nominal diameter (公称直径) D_g/mm	A (Q 以 m^3/s 计)	Nominal diameter (公称直径) D_g/mm	A (Q 以 m^3/s 计)
8	225500000	225.5	125	106.2	400	0.2062
10	32950000	32.95	150	44.95	450	0.1089
15	8809000	8.809	175	18.96	500	0.06222
20	1643000	1.643	200	9.273	600	0.02384
25	436700	0.4367	225	4.822	700	0.01150
32	93860	0.09386	250	2.583	800	0.005665
40	44530	0.04453	275	1.535	900	0.003034
50	11080	0.01108	300	0.9392	1000	0.001736
70	2893	0.002893	325	0.6088	1200	0.0006605
80	1168	0.001168	350	0.4078	1300	0.0004322
100	267.4	0.0002674			1400	0.0002918
125	86.23	0.00008623				
150	33.95	0.00003395				

Table 7-3 Specific Resistance A of Cast-Iron Pipe (s^2/m^6)

表 7-3 铸铁管的比阻 A 值   单位：s^2/m^6

Inner diameter (内径)/mm	A (Q 以 m^3/s 计)	Inner diameter (内径)/mm	A (Q 以 m^3/s 计)	Inner diameter (内径)/mm	A (Q 以 m^3/s 计)
50	15190	250	2.752	600	0.02602
75	1709	300	1.025	700	0.01150
100	365.3	350	0.4529	800	0.005665
125	110.8	400	0.2232	900	0.003034
150	41.85	450	0.1195	1000	0.001736
200	9.029	500	0.06839		

Let
$$S = Al = \frac{8\lambda l}{g\pi^2 d^5} \tag{7-4}$$

So
$$H = SQ^2 \tag{7-5}$$

where S— resistance coefficient，s^2/m^5.

7.1.2 长管

水头损失中绝大部分为沿程损失，其局部损失和流速水头损失相对可以忽略的管路称为长管。

如图 7-2 所示，在断面 1—1 与 2—2 之间应用能量方程可得

$$z_1+\frac{p_1}{\gamma}+\frac{\alpha_1 v_1^2}{2g}=z_2+\frac{p_2}{\gamma}+\frac{\alpha_2 v_2^2}{2g}+h_w$$

又

$$z_1-z_2=H$$

$$\frac{p_1}{\gamma}=\frac{p_2}{\gamma}=0$$

$$\frac{\alpha_1 v_1^2}{2g}=\frac{\alpha_2 v_2^2}{2g}=0$$

$$h_w=h_f$$

代入上述方程，我们可以得到

$$H=h_f=\lambda\frac{l}{d}\frac{v^2}{2g}$$

将 $v=\frac{4Q}{\pi d^2}$ 代入上式得 $H=\frac{8\lambda}{g\pi^2 d^5}lQ^2$。

令

$$A=\frac{8\lambda}{g\pi^2 d^5} \quad (A \text{ 称为比阻}) \tag{7-1}$$

则

$$H=AlQ^2 \tag{7-2}$$

式中，比阻 A 是单位流量通过单位长度管道所需的水头，取决于 λ 和管径 d。

比阻 A 的计算可采用前面所介绍的求 λ 的计算方法。对于旧钢管、旧铸铁管也可以采用舍维列夫公式。

阻力平方区（$v\geqslant 1.2\text{m/s}$）

$$\left.\begin{array}{l} A=\dfrac{0.001736}{d^{5.3}} \\[2mm] \text{过渡区}（v<1.2\text{m/s}） \\[2mm] A'=0.852\left(1+\dfrac{0.867}{v}\right)^{0.3}\left(\dfrac{0.001736}{d^{5.3}}\right)=kA \end{array}\right\} \tag{7-3}$$

式中，k 为修正系数，$k=0.852\left(1+\dfrac{0.867}{v}\right)^{0.3}$。

当水温为 10℃时，各种流速下的 k 值列于表 7-1 中，钢管的比阻 A 及铸铁管的比阻 A 根据式(7-3) 计算所得的值列于表 7-2 和表 7-3 中。

设

$$S=Al=\frac{8\lambda l}{g\pi^2 d^5} \tag{7-4}$$

则

$$H=SQ^2 \tag{7-5}$$

式中，S 为阻抗系数，s^2/m^5。

【Sample Problem 7-2】 As shown in Figure 7-3, there is a water tower to be used to supply water for a factory. A cast-iron pipe is used, and the length and diameter is 2500m and 400mm, respectively. Given that the land elevation of water tower ∇_1 is 61m and the height from ground to water surface in the water tower $H_1=18$m; the land elevation of the factory ∇_2 is 45m and the free head required at pipe ends $H_2=25$m. Try to calculate the flux flowing across the pipeline.

Solution: Based on the sea level (datum), writing energy equation from water surface of the tower to the section of pipe end.

$$(H_1+\nabla_1)+0+0=\nabla_2+H_2+0+h_f$$

Thus
$$h_f=(H_1+\nabla_1)-(H_2+\nabla_2)$$

The head H at pipe end is

$$H=h_f=(61+18)-(45+25)=9\text{m}$$

From Table 7-3, specific resistance A of the cast-iron pipe with a diameter of 400mm is $0.2232\text{s}^2/\text{m}^6$. Substituting it to Equation (7-2) yields

$$Q=\sqrt{\frac{H}{Al}}=\sqrt{\frac{9}{0.2232\times 2500}}=0.127\text{m}^3/\text{s}$$

Checking the velocity of the resistance zone

$$v=\frac{4Q}{\pi d^2}=\frac{4\times 0.127}{\pi\times (0.4)^2}=1.01\text{m/s}<1.2\text{m/s}$$

The flow is transition zone of turbulent flow, so the specific resistance must be corrected. From Table 7-1, when $v=1\text{m/s}$, correction coefficient $k=1.03$. Thus, the flux flowing across the pipeline is

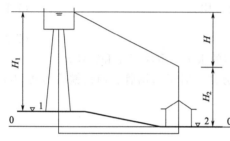

Figure 7-3　Calculation of the Height of Water Tower

图 7-3　水塔高度计算

$$Q=\sqrt{\frac{H}{kAl}}=\sqrt{\frac{9}{1.03\times 0.2232\times 2500}}=0.125\text{m}^3/\text{s}$$

【例题 7-2】 由水塔向工厂供水，如图 7-3 所示，采用铸铁管，管长 2500m，管径 400mm。水塔处地形标高 ∇_1 为 61m，水塔水面距地面高度 $H_1=18$m，工厂地形标高 ∇_2 为 45m，管路末端需要的自由水头 $H_2=25$m，求通过管路的流量。

解： 以海拔水平面为基准面，在水塔水面与管路末端间列出长管的能量方程。

$$(H_1+\nabla_1)+0+0=\nabla_2+H_2+0+h_f$$

故
$$h_f=(H_1+\nabla_1)-(H_2+\nabla_2)$$

管末端作用水头 H 为

$$H=h_f=(61+18)-(45+25)=9\text{m}$$

由表 7-3 查得 400mm 铸铁管比阻 A 为 $0.2232\text{s}^2/\text{m}^6$，代入式(7-2) 得

$$Q=\sqrt{\frac{H}{Al}}=\sqrt{\frac{9}{0.2232\times 2500}}=0.127\text{m}^3/\text{s}$$

验算阻力区

$$v=\frac{4Q}{\pi d^2}=\frac{4\times 0.127}{\pi\times (0.4)^2}=1.01\text{m/s}<1.2\text{m/s}$$

该区域属于过渡区，比阻需要修正，由表 7-1 查得 $v=1\text{m/s}$ 时，$k=1.03$。修正后流量为

$$Q=\sqrt{\frac{H}{kAl}}$$
$$=\sqrt{\frac{9}{1.03\times 0.2232\times 2500}}$$
$$=0.125\text{m}^3/\text{s}$$

7.2 Multiple-Pipe Systems

7.2.1 Pipes in Series

For pipes in series, as shown in Figure 7-4, conditions must satisfy the continuity and energy equation, namely

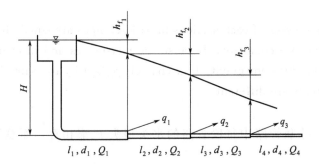

Figure 7-4　Pipes in Series
图 7-4　串联管道

$$Q_i = q_i + Q_{i+1} \tag{7-6}$$

$$H = \sum_{i=1}^{n} h_{fi} = \sum_{i=1}^{n} A_i l_i Q_i^2 = \sum_{i=1}^{n} S_i Q_i^2 \tag{7-7}$$

7.2 复杂管道

7.2.1 串联管道

如图 7-4 所示，串联管道必须满足连续性方程和能量方程，即

$$Q_i = q_i + Q_{i+1} \tag{7-6}$$

$$H = \sum_{i=1}^{n} h_{fi} = \sum_{i=1}^{n} A_i l_i Q_i^2 = \sum_{i=1}^{n} S_i Q_i^2 \tag{7-7}$$

7.2.2 Pipes in Parallel

As shown in Figure 7-5, in the pipes in parallel, the head loss of each section h is equal, but the whole discharge is the summation of each section discharge, namely

$$Q = Q_1 + Q_2 + Q_3 \tag{7-8}$$

$$h_{f_{AB}} = h_{f_1} = h_{f_2} = h_{f_3} \tag{7-9}$$

$$Q = \sqrt{\frac{h_{f_{AB}}}{S_p}},\ Q_1 = \sqrt{\frac{h_{f_{AB}}}{S_1}},\ Q_2 = \sqrt{\frac{h_{f_{AB}}}{S_2}},\ Q_3 = \sqrt{\frac{h_{f_{AB}}}{S_3}}$$

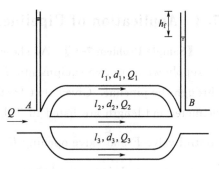

Figure 7-5　Pipes in Parallel
图 7-5　并联管路

Substitute the two formulas above into Equation (7-8), and the result is

$$Q=\sqrt{\frac{h_{f_{AB}}}{S_1}}+\sqrt{\frac{h_{f_{AB}}}{S_2}}+\sqrt{\frac{h_{f_{AB}}}{S_3}}$$

Namely
$$\frac{1}{\sqrt{S_p}}=\frac{1}{\sqrt{S_1}}+\frac{1}{\sqrt{S_2}}+\frac{1}{\sqrt{S_3}} \tag{7-10}$$

The reciprocal of the square root of resistance coefficient S_p of whole resistance in parallel pipes is equal to the summation of reciprocal of square root of resistance coefficient of each section.

Attention: Head losses of each sections in parallel pipes are equal, but it doesn't mean that the energy losses are equal too. Because the resistances and discharges in sections are different, multiplying different gravity flux (namely $\rho g Q_i$) by the same head loss, the power losses in the sections are different.

7.2.2 并联管道

如图 7-5 所示，并联管路中每段管路的水头损失 h 都相等，而总流量为各段流量之和，即

$$Q=Q_1+Q_2+Q_3 \tag{7-8}$$
$$h_{f_{AB}}=h_{f_1}=h_{f_2}=h_{f_3} \tag{7-9}$$

$$Q=\sqrt{\frac{h_{f_{AB}}}{S_p}},\ Q_1=\sqrt{\frac{h_{f_{AB}}}{S_1}},\ Q_2=\sqrt{\frac{h_{f_{AB}}}{S_2}},\ Q_3=\sqrt{\frac{h_{f_{AB}}}{S_3}}$$

代入式(7-8) 中得

$$Q=\sqrt{\frac{h_{f_{AB}}}{S_1}}+\sqrt{\frac{h_{f_{AB}}}{S_2}}+\sqrt{\frac{h_{f_{AB}}}{S_3}}$$

即
$$\frac{1}{\sqrt{S_p}}=\frac{1}{\sqrt{S_1}}+\frac{1}{\sqrt{S_2}}+\frac{1}{\sqrt{S_3}} \tag{7-10}$$

并联管路总阻抗系数 S 的平方根的倒数等于各段管路阻抗系数平方根的倒数之和。

注意：并联管路各段上的水头损失相等并不意味着它们的能量损失也相等，因为各段阻力不同，流量也不同，以同样的水头损失乘以不同的重力流量（即 $\rho g Q_i$）所得到的各段功率损失是不同的。

7.3 Application of Pipelines

【**Sample Problem 7-3**】 As shown in Figure 7-6, use a water pump of which lift is 100m to supply water to two equipments G and H in the workshop where $h_G=40$m and $h_H=60$m through the pipeline. Given that friction resistance coefficient in all pipes is $\lambda=0.024$; the diameter and length are listed in Table 7-4; the relation of the resistance coefficient S_F of control valve F and valve opening K（%）is $S_F=\left(\dfrac{4000}{K}\right)^2$; and the other minor resistances can be ignored. It is required that we can adjust the opening of control valve F to assure that the quantities of water supply of the two equipments G and H are equal. At this time try to calculate:

① Pressure on point E;
② Discharge Q of water pump and supply q of each equipment;
③ Opening K (%) of the control valve.

Table 7-4 The Sizes of Sections
表 7-4 各管段的尺寸

Section (管段)	Length (长度)l/m	Diameter (直径)d/m	Section (管段)	Length (长度)l/m	Diameter (直径)d/m
ABD	30	0.1	EF	15	0.1
ACD	30	0.125	EH	30	0.1
DE	60	0.15			

Solution: Draw out the simplified figure of pipeline to supply water. As shown in Figure 7-7, this is a question about series and parallel of long pipe.

From Equation (7-4), we obtain

$$S=\frac{8\lambda l}{\pi^2 g d^5}=0.0826\lambda\frac{l}{d^5}$$

Substitute data into it and get the resistance coefficient at each section:

$$S_B=5947,\ S_C=1949,\ S_{DE}=1566$$

$$S_{EF}=2974,\ S_{FG}=\left(\frac{4000}{K}\right)^2,\ S_{EH}=5947$$

At first we consider the parallel question of ABD and ACD. Using Equation (7-10), we get

$$\frac{1}{\sqrt{S_{AD}}}=\frac{1}{\sqrt{S_B}}+\frac{1}{\sqrt{S_C}}$$

thus

$$S_{AD}=\frac{S_B S_C}{(\sqrt{S_B}+\sqrt{S_C})^2}=\frac{5947\times 1949}{(\sqrt{5947}+\sqrt{1949})^2}=788$$

Then we consider series question of AD and DE, using Equation (7-7), we get

$$S_{AE}=S_{AD}+S_{DE}=788+1566=2354$$

Assume the head on point A, E and H are h_A, h_E and h_H, respectively. Given that $h_A=100$m, and $h_H=60$m. According to Equation (7-5), the pipeline specialities of section AE and EH are

$$h_A-h_E=S_{AE}Q^2$$

and

$$h_E-h_H=S_{EH}\left(\frac{Q}{2}\right)^2$$

Substituting data into them yields

$$100-h_E=2354Q^2$$

and

$$h_E-60=5947\left(\frac{Q}{2}\right)^2$$

So $h_E=76.71$m, $Q=0.106$m³/s

Pressure on point E

$$p_E=\rho g h_E=9800\times 76.71=751758\text{Pa}$$
$$=7.52\times 10^5\text{Pa}$$

Quantity of water supply to G or H

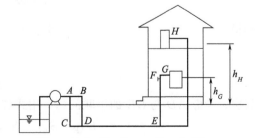

Figure 7-6 Pipeline to Supply Water
图 7-6 供水管路

$$q = \frac{Q}{2} = 0.053 \text{m}^3/\text{s} = 53\text{L/s}$$

Finally, solve the parallel question of EF and valve F, according to Equation (7-7)

$$h_E - h_G = (S_{EF} + S_{FG})\left(\frac{Q}{2}\right)^2$$

Substituting the data

$$76.71 - 40 = \left[2974 + \left(\frac{4000}{K}\right)^2\right]\left(\frac{0.106}{2}\right)^2$$

$$K = 0.398 = 39.8\%$$

Namely, adjusting control valve to this opening can assure the equal quantity of water supply. The quantities of water supply are $q = 53\text{L/s}$.

7.3 管道的应用

【例题 7-3】 如图 7-6 所示，用扬程为 100m 的水泵向车间中位于 $h_G = 40\text{m}$、$h_H = 60\text{m}$ 处的 G、H 两台设备供水。已知所有管段上的沿程阻力系数均为 $\lambda = 0.024$，各管段的长度和直径列于表 7-4 中，调节阀 F 的阻抗系数 S_F 与阀的开度 K（%）的关系是 $S_F = \left(\frac{4000}{K}\right)^2$，忽略其他一切局部阻力。要求调整 F 阀的开度以保证两台设备的供水量完全相等，试求此时：

① E 点处的压强 p_E。
② 水泵的流量 Q 与每台设备的供水量 q。
③ 调节阀的开度 K（%）。

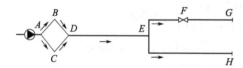

Figure 7-7 Simplified Pipelines
图 7-7 简化管路

解：绘出供水管路的简化图。如图 7-7 所示，这是长管的串并联问题。根据题意，由式(7-4) 得

$$S = \frac{8\lambda l}{\pi^2 g d^5} = 0.0826\lambda \frac{l}{d^5}$$

将数据代入，得出各管段的阻力综合参数为

$$S_B = 5947, \; S_C = 1949, \; S_{DE} = 1566$$

$$S_{EF} = 2974, \; S_{FG} = \left(\frac{4000}{K}\right)^2, \; S_{EH} = 5947$$

首先，考虑 ABD 与 ACD 的并联问题，由式(7-10) 得

$$\frac{1}{\sqrt{S_{AD}}} = \frac{1}{\sqrt{S_B}} + \frac{1}{\sqrt{S_C}}$$

解得

$$S_{AD} = \frac{S_B S_C}{(\sqrt{S_B} + \sqrt{S_C})^2}$$

$$= \frac{5947 \times 1949}{(\sqrt{5947} + \sqrt{1949})^2} = 788$$

其次，考虑 AD 与 DE 的串联问题，用式(7-7) 得

$$S_{AE} = S_{AD} + S_{DE} = 788 + 1566 = 2354$$

设 A、E、H 各点的水头为 h_A、h_E、h_H，已知 $h_A = 100\text{m}$、$h_H = 60\text{m}$，于是根据式(7-5)，可以分别列出 AE 段与 EH 段的管路特性为

$$h_A - h_E = S_{AE}Q^2, \quad h_E - h_H = S_{EH}\left(\frac{Q}{2}\right)^2$$

将数据代入，有

$$100 - h_E = 2354Q^2, \quad h_E - 60 = 5947\left(\frac{Q}{2}\right)^2$$

联立解出

$$h_E = 76.71\text{m}, \quad Q = 0.106\text{m}^3/\text{s}$$

E 点压强 $\quad p_E = \rho g h_E = 9800 \times 76.71 = 751758\text{Pa} = 7.52 \times 10^5 \text{Pa}$

每台设备的供水量为 $\quad q = \dfrac{Q}{2} = 0.053 \text{ m}^3/\text{s} = 53\text{L}/\text{s}$

最后解决 EF 与阀 F 的串联问题，根据式(7-7)

$$h_E - h_G = (S_{EF} + S_{FG})\left(\frac{Q}{2}\right)^2$$

将已知数据代入

$$76.71 - 40 = \left[2974 + \left(\frac{4000}{K}\right)^2\right]\left(\frac{0.106}{2}\right)^2$$

由此解出 $K = 0.398 = 39.8\%$，即将调节阀开到这样的开度，可以保证两台设备的供水量相等，都是 $q = 53\text{L/s}$。

【Sample Problem 7-4】 As shown in Figure 7-8, there are two stand pipes for heating in a two-layer storied building. Given that the diameter and length of section 1 are 20mm and 20m, respectively. The diameter and length of section 2 are 20mm and 10m, respectively. $\Sigma\zeta_1 = 15$, $\Sigma\zeta_2 = 15$, the friction resistance coefficient $\lambda = 0.025$ for all pipes, and the discharge in trunk pipe $Q = 1 \times 10^{-3} \text{m}^3/\text{s}$. Try to calculate the discharges of the two stand pipes Q_1 and Q_2.

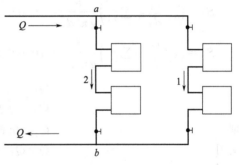

Figure 7-8 Layout of the Stand Pipes for Heating

图 7-8 供暖立管布置

Solution: From Figure 7-8, we can know section 1 and 2 are in parallel between point a and b.

$$S_1 Q_1^2 = S_2 Q_2^2$$

so

$$\frac{Q_1}{Q_2} = \sqrt{\frac{S_2}{S_1}}$$

$$S_1 = \left(\lambda_1 \frac{l_1}{d_1} + \Sigma\zeta_1\right)\frac{8}{g\pi^2 d^4}$$

$$= \left(0.025 \times \frac{20}{0.02} + 15\right)\frac{8}{3.14^2 \times 0.02^4 \times 9.8}$$

$$= 2.07 \times 10^7 \text{s}^2/\text{m}^5$$

$$S_2 = \left(\lambda_2 \frac{l_2}{d_2} + \Sigma\zeta_2\right)\frac{8}{g\pi^2 d^4}$$

$$= \left(0.025 \times \frac{10}{0.02} + 15\right)\frac{8}{3.14^2 \times 0.02^4 \times 9.8}$$

$$= 1.42 \times 10^7 \text{s}^2/\text{m}^5$$

$$\frac{Q_1}{Q_2} = \sqrt{\frac{1.42 \times 10^7}{2.07 \times 10^7}} = 0.828$$

Namely $\quad Q_1 = 0.828 Q_2$

As $\quad Q = Q_1 + Q_2 = 0.828 Q_2 + Q_2 = 1.828 Q_2$

$$Q_2 = \frac{1}{1.828} Q = 0.55 \times 10^{-3} \text{m}^3/\text{s}$$

$$Q_1 = 0.828 Q_2 = 0.45 \times 10^{-3} \text{m}^3/\text{s}$$

From Sample Problem 7-4, we can know pipes in series or in parallel also can be calculated as short pipes, except that the effects of minor resistance coefficient on specific resistance A or resistance coefficient S should be calculated.

【例题 7-4】 某两层楼的供暖立管，管段 1 的直径为 20mm，总长度为 20m，$\Sigma \zeta_1 = 15$。管段 2 的直径为 20mm，总长度为 10m，$\Sigma \zeta_2 = 15$。管路的 $\lambda = 0.025$，干管中的流量 $Q = 1 \times 10^{-3} \text{m}^3/\text{s}$，求 Q_1 和 Q_2。

解：从图 7-8 可知，节点 a、b 间并联有 1、2 两管段。

由 $S_1 Q_1^2 = S_2 Q_2^2$ 得 $\dfrac{Q_1}{Q_2} = \sqrt{\dfrac{S_2}{S_1}}$，计算 S_1 和 S_2：

$$S_1 = \left(\lambda_1 \frac{l_1}{d_1} + \Sigma \zeta_1 \right) \frac{8}{g \pi^2 d^4} = \left(0.025 \times \frac{20}{0.02} + 15 \right) \frac{8}{3.14^2 \times 0.02^4 \times 9.8} = 2.07 \times 10^7 \text{s}^2/\text{m}^5$$

$$S_2 = \left(\lambda_2 \frac{l_2}{d_2} + \Sigma \zeta_2 \right) \frac{8}{g \pi^2 d^4} = \left(0.025 \times \frac{10}{0.02} + 15 \right) \frac{8}{3.14^2 \times 0.02^4 \times 9.8} = 1.42 \times 10^7 \text{s}^2/\text{m}^5$$

$$\frac{Q_1}{Q_2} = \sqrt{\frac{1.42 \times 10^7}{2.07 \times 10^7}} = 0.828$$

则 $\quad Q_1 = 0.828 Q_2$

又因 $\quad Q = Q_1 + Q_2 = 0.828 Q_2 + Q_2 = 1.828 Q_2$

$$Q_2 = \frac{1}{1.828} Q = 0.55 \times 10^{-3} \text{m}^3/\text{s}$$

$$Q_1 = 0.828 Q_2 = 0.45 \times 10^{-3} \text{m}^3/\text{s}$$

从例题 7-4 可以看出，串联、并联管路也可以按短管计算，只需在比阻 A 或阻抗 S 中考虑局部阻力系数的影响。

Exercises 习题

7-1 A smooth pipe consists of 120m of 200-mm-diameter pipe followed by 100m of 600-mm-diameter pipe, with an abrupt change of cross section at the junction. The entrance is flush and the discharge is submerged. If it carries water at 20℃, with a velocity of 6.0m/s in the smaller pipe, what is the total head loss?

7-1 一光滑管由一根长 120m、直径为 200mm 的水管和一根长 100m、直径为 600mm 的水管组成，在连接点变径。入口处是齐平的，淹没排放。如果在 20℃的水温下，在小口径的水管

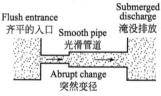

Figure for Exercise 7-1
习题 7-1 图

中以 6.0m/s 的速度输水，那么总水头损失为多少？

7-2 A pipeline with a length of 200m discharges freely at a point 30m lower than the water surface at intake. The pipe intake projects into the reservoir. The first 150m is of 300mm diameter, and the remaining 50m is of 200mm diameter. Assuming $\lambda = 0.06$, the junction point C of the two sizes of pipe is 40m below the intake water surface level. Assume a sudden contraction at C. ① Find the rate of discharge; ② Find the pressure head just upstream of C; ③ Find the pressure head just downstream of C.

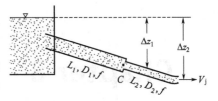

Figure for Exercise 7-2

习题 7-2 图

7-2 长 200m 的管道在比入口处的水平面低 30m 处自由排放，管道的入口插入蓄水池内。起始段为长 150m、直径 300mm 的管子，其余管道长 50m、直径为 200mm。假设 $\lambda=0.06$，两管道连接点 C 比入口处的水平面低 40m，且在 C 点突然收缩。① 求排放量；② 求 C 点上游的压强水头；③ 求 C 点下游的压强水头。

7-3 In Figure, ① if the total head loss in Branch C is 2.87m, what is the total head loss in Branch D? ② if the initial flow in Branch D is 2.0m³/s, what will be the flow in Branch D if the flow from A to B is doubled?

7-3 在图中，① 假如支流 C 的总水头损失为 2.87m，那么支流 D 的总水头损失为多少？② 假如支流 D 的初始流量为 2.0m³/s，如果从 A 到 B 的流量增加一倍，那么支流 D 中的流量为多少？

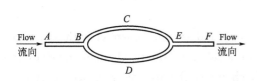

Figure for Exercise 7-3

习题 7-3 图

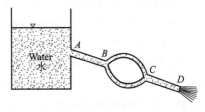

Figure for Exercise 7-4

习题 7-4 图

7-4 In Figure pipe AB is 500m long, of 300mm diameter, with $\lambda = 0.035$; pipe BC (upper) is 500m long, of 200mm diameter, with $\lambda = 0.025$; pipe BC (lower) is 400m long, of 250mm diameter, with $\lambda = 0.030$; and pipe CD is 900m long, of 320mm diameter, with $\lambda = 0.020$. The elevations are: reservoir water surface $=150$m, $A=100$m, $B=60$m, $C=40$m, $D=20$m.

There is free discharge to the atmosphere at D. Neglecting velocity heads.

① Compute the flow in each pipe;

② Determine the pressure at B and C.

7-4 如图所示，管道 AB 长 500m，直径 300mm，$\lambda=0.035$；管道 BC 上面的那段长 500m，直径 200mm，$\lambda=0.025$，下面的那段长 400m，直径 250mm，$\lambda=0.030$；管道 CD 长 900m，直径 320mm，$\lambda=0.020$。高程分别为：蓄水池水平面高程 150m，$A=100$m，$B=60$m，$C=40$m，$D=20$m。在 D 点自由排放，忽略流速水头。

① 计算每段水管的流量；

② 求 B 点和 C 点的压强。

Chapter 8　Flow in Open Channels
第 8 章　明　渠　流

8.1　Open Channel

An open channel is one in which the stream has a free surface subjected only to atmospheric pressure. According to ways of formation, open channels are classified as natural open channels (natural streams and rivers) and artificial open channels (aqueducts, sewers, canals and pipelines flowing not completely full etc.).

Different from pressure flow, open-channel flow has a free surface subjected only to atmospheric pressure, and the gage pressure is equal to zero. So we often refer to open-channel flow as nonpressure flow.

The streams in an open channel are divided into steady flow and unsteady flow according to motional conditions in the streams with respect to time. In steady flow, when the streamlines are parallel line, we say the flow is uniform flow. Otherwise, the flow is nonuniform flow.

The types of open channels are as follows:

① Prismatic and nonprismatic channels. The long and straight channel with a changeless shape and size of cross-section along the stream is called prismatic channel, or else, the channel is nonprismatic channel. For the prismatic channel, the area of cross-section A only changes with depth h, but for nonprismatic channel, the area of cross-section A changes with depth h, as well as the location s of cross-section.

② Falling slope, horizontal slope and rising slope channels. Usually the bed of an open channel is a slope. At vertical section, channel bed is an oblique line and slope of the oblique line is called bed slope of open channel i. In general, if $i>0$, the bed slope is called falling slope; if $i=0$, the bed slope is called horizontal slope; and if $i<0$, the bed slope is called rising slope.

Bed slope of open channel i is the ratio of depth difference of bed to corresponding channel length l, namely

$$i=\frac{\Delta z}{l}=\sin\theta \tag{8-1}$$

As shown in Figure 8-1, in the formula above, θ is the inclination between bed and horizontal.

Usually the bed slope is very small ($i \leqslant 0.01$), namely, θ is very small, so we can think that the length of bed along stream l is equal to the length of its horizontal projection l_x. Thus,

$$i = \frac{\Delta z}{l_x} = \tan\theta \tag{8-2}$$

Moreover, when bed slope is very small, we can think that the cross-section has no difference with vertical section in the stream. So the cross-section and depth can be measured along vertical.

8.1 明渠

明渠是一种具有自由表面水流的渠道，根据形成方式分为天然明渠（天然河道）和人工明渠（输水渠、排水渠、运河及未充满水流的管道等）。

明渠流与有压管流不同，它具有自由表面，表面上受大气压强作用，相对压强为零，所以又称为无压流动。明渠水流根据其运动要素是否随时间变化分为恒定流动与非恒定流动，恒定流动又根据流线是否为平行直线分为均匀流动与非均匀流动两类。

明渠的分类如下。

① 棱柱形渠道与非棱柱形渠道。凡是断面形状及尺寸沿程不变的长直渠道称为棱柱形渠道，否则为非棱柱形渠道。前者的过水断面面积 A 仅随水深 h 变化，后者过水断面面积 A 既随水深 h 变化，又随各断面沿程位置 s 而变化。

② 顺坡、平坡和逆坡渠道。明渠底一般是个斜坡。在纵剖面上，渠底便成一条斜直线，这一斜线即渠底线的坡度，称作渠道底坡 i。一般规定，渠底沿程降低的底坡 $i>0$，称为顺坡；渠底水平时 $i=0$，称为平坡；渠底沿程升高时 $i<0$，称为逆坡。

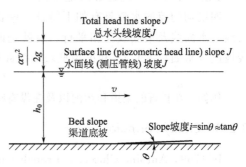

Figure 8-1　Uniform Flow in an Open Channel
图 8-1　明渠均匀流

渠道底坡 i 是指渠底的高差 Δz 与相应渠长 l 的比值，故有

$$i = \frac{\Delta z}{l} = \sin\theta \tag{8-1}$$

式中，θ 为渠底与水平线间的夹角，见图 8-1。

注意，通常土渠的底坡很小（$i \leqslant 0.01$），即 θ 角很小，在实际中可以认为渠道底线沿水流方向的长度 l 和它的水平投影长度 l_x 相等，即

$$i = \frac{\Delta z}{l_x} = \tan\theta \tag{8-2}$$

另外，在渠道底坡微小的情况下，水流的过水断面与铅垂断面可以认为没有差异。因此过水断面和水流深度可沿垂直方向量取。

8.2 Calculation Equation of Uniform Flow in Open Channel

Uniform flow in open channel is a motion with uniform velocity. From force balance law, we know the component of gravity at the stream orientation, namely, the thrust of stream is in equilibrium with the friction force which resists the motion of water. So the conditions of uniform flow in open channel are that both bed slope i and coefficient of roughness

n must be uniform along the stream, furthermore, the bed slope must be a falling slope.

Uniform flow in open channel has some characteristics as follows: both the mean velocity v and depth h are changeless along the stream; and total head line, surface line and bed line are in parallel each other. That is to say, the total head line slope (hydraulic slope) J, piezometric head line slope (surface line slope) J_p and bed slope i equal each other, namely,

$$J = J_p = i \tag{8-3}$$

For example, in long and straight channels and canals, and natural long straight rivers, the flows are uniform flow approximately.

8.2 明渠均匀流的计算公式

明渠均匀流是等速运动，根据静力平衡原理可知，重力在水流方向上的分力即水流运动的推力与阻碍水流运动的摩擦阻力相平衡。因此，明渠均匀流发生的条件是底坡 i 和粗糙率 n 必须沿程不变，而且还必须是顺坡。

明渠均匀流的水流具有如下特征：断面平均流速 v 沿程不变；水深 h 沿程不变；总水头线、水面线及渠底线互相平行。也就是说，其总水头线坡度（水力坡度）J、测压管水头线坡度（水面坡度）J_p 和渠道底坡 i 彼此相等，即

$$J = J_p = i \tag{8-3}$$

例如，在长直的渠道和运河以及在没有障碍的天然顺直河段中，其水流便近乎均匀流动。

8.2.1 Chézy Formula

In 1769, Antoine Chézy, a French engineer, brought forward a calculation equation of uniform flow in open channel, namely, Chézy formula.

$$v = C\sqrt{RJ} \tag{8-4}$$

where　v—mean velocity, m/s;

　　　　R—hydraulic radius, m;

　　　　J—hydraulic slope;

　　　　C—coefficient of velocity, or Chézy coefficient, $m^{1/2}/s$.

For uniform flow in open channel, $J = i$, so Equation (8-4) can be written as the following formula:

$$v = C\sqrt{Ri} \tag{8-5}$$

We can get the formula for flux calculation:

$$Q = Av = AC\sqrt{Ri} = K\sqrt{i} = K\sqrt{J} \tag{8-6}$$

$$K = AC\sqrt{R}$$

where　K — flux modulus, and its unit is the same as the unit of flux;

　　　　A — the area of cross-section with a depth of h in uniform flow in open channel.

The depth h, which corresponds to $K = Q/\sqrt{i}$, is called normal depth and is denoted as h_0 usually.

8.2.1 谢才公式

1769 年，法国工程师谢才（Antoine Chézy）提出了明渠均匀流的计算公式，即谢才公式。

$$v = C\sqrt{RJ} \tag{8-4}$$

式中，v 为平均流速，m/s；R 为水力半径，m；J 为水力坡度；C 为水流流速系数，亦称为谢才系数，$m^{1/2}/s$。

由于在明渠均匀流中，$J=i$，故式(8-4) 可写为

$$v=C\sqrt{Ri} \tag{8-5}$$

则得流量的算式

$$Q=Av=AC\sqrt{Ri}=K\sqrt{i}=K\sqrt{J} \tag{8-6}$$

式中，$K=AC\sqrt{R}$，它的单位与流量相同，称为流量模数；A 为对应明渠均匀流水深 h 的过水断面面积。

对应 $K=Q/\sqrt{i}$ 的水深 h 称为正常水深，通常以 h_0 表示。

8.2.2 Manning Formula

In 1890, Robert Manning, an Irish engineer, brought forward a calculation equation of uniform flow in open channel too, namely, Manning formula.

$$v=\frac{1}{n}R^{2/3}J^{1/2} \tag{8-7}$$

Comparing Chézy formula [Equation (8-4)] with Manning formula [Equation (8-7)], we obtain

$$C=\frac{1}{n}R^{1/6} \tag{8-8}$$

Equation (8-8) shows there is an important relation between Chézy coefficient C and coefficient of roughness n, so C is also called Manning coefficient. Coefficients of roughness on different rough surfaces are listed in Table 8-1.

8.2.2 曼宁公式

1890 年，爱尔兰工程师曼宁（Robert Manning）亦提出了一个明渠均匀流公式，即曼宁公式

$$v=\frac{1}{n}R^{2/3}J^{1/2} \tag{8-7}$$

将谢才公式 [式(8-4)] 与曼宁公式 [式(8-7)] 相比较，便得

$$C=\frac{1}{n}R^{1/6} \tag{8-8}$$

式(8-8) 表明了谢才系数 C 与粗糙系数 n 之间的重要关系，因此也称 C 为曼宁系数。各种不同粗糙面的粗糙系数 n 见表 8-1。

Table 8-1　Coefficients of Roughness on Different Rough Surfaces

表 8-1　各种不同粗糙面的粗糙系数 n

Grades 等级	Channel wall types 槽壁种类	n	$1/n$
1	Enamel coating surfaces; a very fine polished and well-pieced wood board 涂覆珐琅釉质的表面；极精细抛光且拼合良好的木板	0.009	111.1
2	A polished wood board; unadulterated cement plaster finish 抛光的木板；纯粹水泥的粉饰面	0.010	100.0

续表

Grades 等级	Channel wall types 槽壁种类	n	$1/n$
3	Cement plaster finish(fine sand accounted for a third); new pottery clay; well-installed cast iron pipe and steel pipe 水泥(含三分之一细沙)粉饰面,新陶土,安装和接合良好的铸铁管和钢管	0.011	90.9
4	A non-polished and well-pieced wood board; feed pipe with no significant fouling; excellent concrete surface 未抛光而拼合良好的木板;正常情况下内无显著积垢的给水管;极好的混凝土面	0.012	83.3
5	Ashlar masonry; excellent brick masonry; drainage pipe under common circumstance; feed pipe with slight pollution; non-polished wood board with poor piecing quality 料石砌体;极好的砖砌体;正常情况下的排水管;略微污染的给水管;非完全精密拼合的、未抛光的木板	0.013	76.9
6	Contaminated feed pipe and drainage pipe; ordinary brick masonry; concrete surface of channel under common circumstance 污染的给水管和排水管;一般的砖砌体;一般情况下渠道的混凝土面	0.014	71.4
7	Rough brick masonry; non-polished stone masonry; smooth stones; drainage pipe with significant fouling 粗糙的砖砌体;未琢磨的石砌体;安置平整的石块;污垢显著的排水管	0.015	66.7
8	Ordinary stone masonry with good condition; old broken brick masonry; a rough concrete; a smooth cliff bank with extremely good digging quality 状况良好的普通石块砌体;破旧砖砌体;较粗糙的混凝土;光滑的开凿得极好的崖岸	0.017	58.8
9	Channel covered with a thick silt layer; channel made with dense loess and dense pebble and covered with a thin silt layer; rough stone masonry; dry cyclopean masonry; pebble paving surface 覆有坚厚淤泥层的渠槽;用致密黄土和致密卵石做成且为整片淤泥薄层所覆盖的良好渠槽;很粗糙的块石砌体;大块石的干砌体;卵石铺筑面	0.018	55.6
10	Cutting channel in the rock; channel made with loess, dense pebble and dense soil and covered with a thin silt layer(under common circumstance) 岩石中开筑的渠槽;由黄土、致密卵石和致密泥土做成且为淤泥薄层所覆盖的渠槽(正常情况)	0.020	50.0
11	Large sharpstone paving surface; rock channel with ordinary processing surface; channel made with dense clay; channel made with loess, pebble and soil and covered with a partial thin silt layer(some places are fractured); large channel under moderate or above maintenance 尖角的大块乱石铺筑的表面;表面经过普通处理的岩石渠槽;致密黏土渠槽;由黄土、卵石和泥土做成且部分(有些地方断裂的)被淤泥薄层所覆盖的渠槽;受中等以上养护的大型渠槽	0.0225	44.4
12	Large soil channel maintained at moderate level; small well-maintained soil channel; rivers and streams under favorable conditions(water flows freely with no blockage and no remarkable quantity of waterweeds and so on) 受到中等养护的大型土渠;受到良好养护的小型土渠;有利条件(自由流动,无淤塞和显著水草等)下的小河和溪涧	0.025	40.0
13	Large channel maintained below moderate level; small channel maintained at moderate level 中等条件以下的大渠道;中等条件的小渠槽	0.0275	36.4
14	Channels and rivers with poor maintenance(for example, there are waterweeds and free stones in some places or remarkable quantity of tall grass or partial landslide in some places, etc.) 条件较差(例如有些地方有水草和乱石或显著的杂草,有局部的坍坡等)的渠道和小河	0.030	33.3
15	Channels and rivers with very poor maintenance(the cross-section is irregular, or water flow is blocked severely by waterweeds and free stones, etc.) 条件很差的渠道和小河(断面不规则,受到石块和水草的严重阻塞等)	0.035	28.6
16	Channels and rivers with particularly poor maintenance(along the channels and rivers, there are large stones from collapsed cliff, dense roots, deep pool and bank collapses and so on) 条件特别差的渠道和小河(沿河有崩崖的巨石、绵密的树根、深潭、坍岸等)	0.040	25.0

8.2.3 Павповский Formula

In 1925, Н. Н. Павповский, a Soviet hydraulician, brought forward a formula with a variational exponential, and this formula is called Павповский formula.

$$C = \frac{1}{n} R^y \tag{8-9}$$

$$y = 2.5\sqrt{n} - 0.13 - 0.75\sqrt{R}\,(\sqrt{n} - 0.10)$$

where the range of parameters are: $0.1\text{m} \leqslant R \leqslant 3\text{m}$ and $0.011 < n < 0.040$.

【Sample Problem 8-1】 There is a small long straight river with a length of 1km. Some stones are on the riverbed and some aquatic plants are on the sides of bank. Given that the cross-section of the river is a trapezoid and side slope $m = \cot\alpha = 1.5$. The drop, width and depth of bed are 0.5m, 3m and 0.8m, respectively (shown in Figure 8-2). Try to calculate the flux modulus K and flux Q using Manning formula and Павповский formula.

Solution: We can calculate according to the formula $Q = AC\sqrt{Ri} = K\sqrt{i}$. Calculate the bed slope firstly

$$i = \frac{0.5}{1000} = 0.0005$$

The area of cross-section $A = (b + mh)h = (3 + 1.5 \times 0.8) \times 0.8 = 3.36\text{m}^2$.

Wetted perimeter $\chi = b + 2h\sqrt{1+m^2} = 3 + 2 \times 0.8\sqrt{1+1.5^2} = 5.88\text{m}$.

Hydraulic radius $R = \frac{A}{\chi} = \frac{3.36}{5.88} = 0.57\text{m}$.

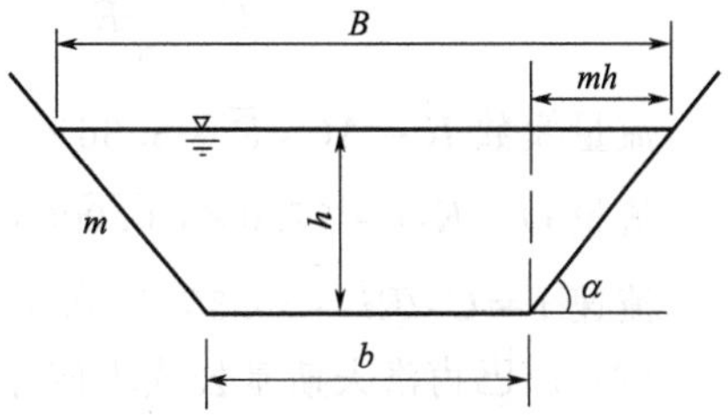

From Table 8-1, coefficient of roughness $n = 0.03$.

(1) Calculation of Chézy coefficient using Manning formula

$$C = \frac{1}{n} R^{1/6} = \frac{1}{0.03}(0.57)^{1/6} = 30.35\text{m}^{1/2}/\text{s}$$

Figure 8-2 A Trapezoidal Cross-Section of an Open Channel

图 8-2 明渠梯形断面

Flux modulus $K = AC\sqrt{R} = 3.36 \times 30.35 \times \sqrt{0.57} = 77.0\text{m}^3/\text{s}$.

Flux $Q = K\sqrt{i} = 77.0 \times \sqrt{0.0005} = 1.72\text{m}^3/\text{s}$.

Velocity $v = C\sqrt{Ri} = 30.35 \times \sqrt{0.57 \times 0.0005} = 0.51\text{m/s}$.

(2) Calculation of Chézy coefficient using Павповский formula

$$C = \frac{1}{n} R^y$$

where $y = 0.26$. So Chézy coefficient $C = 28.8\text{ m}^{\frac{1}{2}}/\text{s}$.

Flux modulus $K = AC\sqrt{R} = 3.36 \times 28.8 \times \sqrt{0.57} = 73.06\text{m}^3/\text{s}$.

Flux $Q = K\sqrt{i} = 73.06 \times \sqrt{0.0005} = 1.63\text{m}^3/\text{s}$.

Velocity $v = C\sqrt{Ri} = 28.8 \times \sqrt{0.57 \times 0.0005} = 0.486\text{m/s}$.

8.2.3 巴甫洛夫斯基公式

1925年苏联水力学家巴甫洛夫斯基（Н. Н. Павповский）提出了一个带有变指数的公式，称为巴甫洛夫斯基公式。

$$C = \frac{1}{n} R^y \tag{8-9}$$

$$y = 2.5\sqrt{n} - 0.13 - 0.75\sqrt{R}(\sqrt{n} - 0.10)$$

此式是在下列数据范围内得到的：$0.1\text{m} \leq R \leq 3\text{m}$ 及 $0.011 < n < 0.040$。

【例题 8-1】 有一段长为1km的顺直小河，河床有乱石，岸边有水草，这段河床的过水断面为梯形，其底部落差为0.5m，底宽3m，水深0.8m，边坡系数 $m = \cot\alpha = 1.5$（图8-2）。试用曼宁公式和巴甫洛夫斯基公式求流量模数 K 和流量 Q。

解：根据基本关系式 $Q = AC\sqrt{Ri} = K\sqrt{i}$ 进行计算，先求渠底坡度

$$i = \frac{0.5}{1000} = 0.0005$$

过水断面面积 $A = (b + mh)h = (3 + 1.5 \times 0.8) \times 0.8 = 3.36 \text{m}^2$。

湿周 $\chi = b + 2h\sqrt{1+m^2} = 3 + 2 \times 0.8\sqrt{1+1.5^2} = 5.88\text{m}$。

水力半径 $R = \frac{A}{\chi} = \frac{3.36}{5.88} = 0.57\text{m}$。

查表8-1得粗糙系数 $n = 0.03$。

（1）用曼宁公式求谢才系数

$$C = \frac{1}{n} R^{1/6} = \frac{1}{0.03}(0.57)^{1/6} = 30.35 \text{m}^{1/2}/\text{s}$$

流量模数 $K = AC\sqrt{R} = 3.36 \times 30.35 \times \sqrt{0.57} = 77.0 \text{m}^3/\text{s}$。

流量 $Q = K\sqrt{i} = 77.0 \times \sqrt{0.0005} = 1.72 \text{m}^3/\text{s}$。

流速 $v = C\sqrt{Ri} = 30.35 \times \sqrt{0.57 \times 0.0005} = 0.51 \text{m/s}$。

（2）用巴甫洛夫斯基公式求谢才系数

$$C = \frac{1}{n} R^y$$

式中，$y = 0.26$。则 $C = 28.8 \text{ m}^{\frac{1}{2}}/\text{s}$。

流量模数 $K = AC\sqrt{R} = 3.36 \times 28.8 \times \sqrt{0.57} = 73.06 \text{m}^3/\text{s}$。

流量 $Q = K\sqrt{i} = 73.06 \times \sqrt{0.0005} = 1.63 \text{m}^3/\text{s}$。

流速 $v = C\sqrt{Ri} = 28.8 \times \sqrt{0.57 \times 0.0005} = 0.486 \text{m/s}$。

8.3 Most Efficient Cross-section and Allowable Velocity

8.3.1 Most Efficient Cross-section

From the discussion above, we can know water transport capability depends on bed slope, coefficient of roughness, and the shape and size of cross-section. During design of open channel, bed slope i is decided by the terrain, and coefficient of roughness n depends on material of channel wall. And then water transport capability Q only depends on the shape and size of cross-section. For a given i, n and A, the shape of cross-section through which the flux is a maximum is called most efficient cross-section.

Due to the basic relation of uniform flow, we get

$$Q = AC\sqrt{Ri} = A\left(\frac{1}{n}R^{1/6}\right)\sqrt{Ri} = \frac{A}{n}R^{2/3}i^{1/2} = \frac{\sqrt{i}}{n} \times \frac{A^{5/3}}{\chi^{2/3}}$$

From the formula above we can know that when i, n and A are given, hydraulic radius R is a maximum, i.e., the cross-section with a minimal wetted perimeter χ can flow across the maximal flux.

Of all geometric figures, the circle has the least perimeter for a given area A. So shape of cross-section of pipeline is a circle usually. For open channel, the shape of cross-section is semicircle, but the semicircle open channel is difficult to construct, so semicircle open channel is only built of reinforced concrete or steel wire netting concrete. Canals excavated in earth usually have a trapezoidal cross-section. Thereinto, a trapezoidal cross-section equivalent to half of a regular hexagon closes to the most efficient semicircle cross-section mostly. The side slope of this cross-section

$$m = \cot\alpha = \cot 60° = 0.557$$

Side slopes of cross-sections in the form of trapezoids m in various soils are shown in Table 8-2. In fact, we often determine side slope m according to the soil on channel surface or the characteristics of protective covering firstly, and use them to calculate the most efficient semicircle cross-section.

8.3 明渠水力最优断面和允许流速

8.3.1 水力最优断面

从以上讨论可知，明渠均匀流输水能力的大小取决于渠道底坡、粗糙系数以及过水断面的形状和尺寸。在设计渠道时，底坡 i 一般随地形条件而定，粗糙系数 n 取决于渠壁的材料，于是，渠道输水能力 Q 只取决于断面大小和形状。当 i、n 及 A 大小一定时，使渠道所通过的流量最大的断面形状称为水力最优断面。

从均匀流的基本关系式得

$$Q = AC\sqrt{Ri} = A\left(\frac{1}{n}R^{1/6}\right)\sqrt{Ri} = \frac{A}{n}R^{2/3}i^{1/2} = \frac{\sqrt{i}}{n} \times \frac{A^{5/3}}{\chi^{2/3}}$$

从上式可以看出，当 i、n 及 A 给定时，水力半径 R 最大即湿周 χ 最小的断面能通过最大的流量。

面积 A 为定值，边界最小的几何图形是圆形，因此管路断面形状通常为圆形，对于明渠则为半圆形，但半圆形施工困难，只在钢筋混凝土或钢丝网水泥材料为断面时采用。土壤开挖的渠道一般都用梯形断面，其中最接近水力最优断面半圆形的是一种相当于半个正六边形的梯形断面。这种梯形所要求的边坡系数

$$m = \cot\alpha = \cot 60° = 0.557$$

对于不同种类的土壤，梯形过水断面边坡系数 m 见表 8-2。实际工程中，常常先根据渠面土壤或护面性质来确定它的边坡系数 m，并以此算出水力最优的梯形过水断面。

Table 8-2　Side Slopes of Cross-Sections in the Form of Trapezoids m

表 8-2　梯形过水断面的边坡系数 m

Soil types 土壤种类	Side slopes m 边坡系数 m	Soil types 土壤种类	Side slopes m 边坡系数 m
Fine sand 细粒砂土	3.0~3.5	Heavy soil, compacted loess, clay 重土壤、密实黄土、普通黏土	1.0~1.5
Sandy loam, loose soil 砂壤土或松散土壤	2.0~2.5	Dense heavy clay 密实重黏土	1.0
Dense sandy loam, light clay loam 密实砂壤土、轻黏土壤	1.5~2.0	Rocks of different hardness 各种不同硬度的岩石	0.5~1.0
Gravel, sandy gravel soil 砾石、砂砾石土	1.5		

8.3.2　Hydraulic Optimum Condition of a Trapezoid Cross-Section

Consider the trapezoid (Figure 8-2), with bed width b, depth h and side slope m. So the area of cross-section $A = (b+mh)h$. Hence we obtain

$$b = A/h - mh$$

and
$$\chi = b + 2h\sqrt{1+m^2} = \frac{A}{h} - mh + 2h\sqrt{1+m^2} \tag{8-10}$$

Differentiate χ with respect to h and set the result equal to zero.

$$\frac{d\chi}{dh} = -\frac{A}{h^2} - m + 2\sqrt{1+m^2} = 0 \tag{8-11}$$

Differentiating χ with respect to h again yields

$$\frac{d^2\chi}{dh^2} = 2\frac{A}{h^3} > 0$$

So we can find $\chi_{\min}$. By substituting $A = (b+mh)h$ into Equation (8-11) and resolving it, we can obtain the hydraulic optimum condition denoted by $\beta = b/h$ for a trapezoid cross-section.

$$\beta = \frac{b}{h} = 2(\sqrt{1+m^2} - m) \tag{8-12}$$

$$b = (2\sqrt{1+m^2} - 2m)h$$

From this, we can know the ratio of width to depth β of most efficient cross-section is a function of only side slope m, and the β values from Equation (8-12) with various m values are listed in Table 8-3.

From Equation (8-12), we can also draw a conclusion that for most efficient cross-section the hydraulic radius R is half of depth h in spite of any side slope m. The deriving process is as follows:

$$R = \frac{A}{\chi} = \frac{(b+mh)h}{b+2h\sqrt{1+m^2}}$$

Substituting Equation (8-12) into equation above yields

$$R_{\max} = \frac{h}{2} \tag{8-13}$$

When $m = 0$, Equation (8-12) changes to

$$\beta = 2, \text{ namely, } b = 2h \tag{8-14}$$

This shows bed width b is twice of depth h for a most efficient rectangle cross-section.

For a small channel, its most efficient cross-section is very close to its economic optimum cross-section. While for a big channel, the most efficient cross-section is narrow and deep, and it is difficult to build, thus it is not an economic optimum cross-section. As a result, when we design a channel, all kinds of factors need to be considered. Here the hydraulic optimum condition is one of considering factors.

8.3.2 边坡系数已定前提下的梯形断面水力最优条件

设明渠梯形过水断面（图 8-2）的底宽为 b，水深为 h，边坡系数为 m，于是过水断面面积为

$$A = (b + mh)h$$

解得

$$b = A/h - mh$$

湿周

$$\chi = b + 2h\sqrt{1+m^2} = \frac{A}{h} - mh + 2h\sqrt{1+m^2} \tag{8-10}$$

对式(8-10)求导，并令其导数为零

$$\frac{d\chi}{dh} = -\frac{A}{h^2} - m + 2\sqrt{1+m^2} = 0 \tag{8-11}$$

再求二阶导数，得

$$\frac{d^2\chi}{dh^2} = 2\frac{A}{h^3} > 0$$

故有 χ_{min} 存在。现解式(8-11)，并将 $A = (b+mh)h$ 代入，便得到以宽深比 $\beta = b/h$ 表示的梯形过水断面的水力最优条件：

$$\beta = \frac{b}{h} = 2(\sqrt{1+m^2} - m) \tag{8-12}$$

$$b = 2(\sqrt{1+m^2} - 2m)h$$

由此可见，水力最优断面的宽深比 β 仅是边坡系数 m 的函数，根据式(8-12)可列出不同 m 时的 β 值，见表 8-3。

Table 8-3 Ratio of Width to Depth β of the Most Efficient Cross-Section

表 8-3 水力最优断面的宽深比 β

$m = \cot\alpha$	0	0.25	0.50	0.75	1.00	1.25	1.50	1.75	2.00	3.00	3.5
$\beta = b/h$	2.00	1.56	1.24	1.00	0.83	0.70	0.61	0.53	0.47	0.32	0.28

从式(8-12)出发，还可以引出一个结论，在任何边坡系数 m 下，水力最优梯形断面的水力半径 R 为水深 h 的一半。现推导如下：

$$R = \frac{A}{\chi} = \frac{(b+mh)h}{b + 2h\sqrt{1+m^2}}$$

代入水力最优断面条件式(8-12)得

$$R_{max} = \frac{h}{2} \tag{8-13}$$

当 $m = 0$ 时，式(8-12)变为

$$\beta = 2, \text{ 即 } b = 2h \tag{8-14}$$

说明水力最优矩形断面的底宽 b 为水深 h 的 2 倍。

对于小型渠道，它的水力最优断面和其经济合理断面比较接近。对于大型渠道，水力最优断面窄而深，不利于施工，因而不是最经济合理的断面。因而在做渠道设计时，需要综合各方面的因素来考虑，这里所提出的水力最优条件便是一种考虑因素。

8.3.3 Allowable Velocity of Open Channel

Besides hydraulic optimum conditions and costs, the design velocity of open channel should be neither so great as to scour channel bed, nor so small as to silt mud and sand suspending in water. We should choose a design velocity between non-scouring velocity and non-silting velocity. Namely,

$$v_{max} > v > v_{min}$$

where v_{max} — the maximum velocity and is called non-scouring velocity for short;

v_{min} — the minimum velocity and is called non-silting velocity for short.

Non-scouring velocity of channel depends on characteristics of soil or lined material. Non-scouring velocities for all kinds of open channels are listed in Table 8-4, and they can be used to design open channels.

Non-silting velocity v_{min} of channel can be chosen according to the following criterion:

① Preventing plant from growing, $v_{min} = 0.6 \text{m/s}$;
② Preventing mud from silting, $v_{min} = 0.2 \text{m/s}$;
③ Preventing sand from depositing, $v_{min} = 0.4 \text{m/s}$.

8.3.3 渠道的允许流速

除考虑水力最优条件及经济因素外，还应使渠道设计流速不应大到使渠床受到冲刷，也不可小到使水中悬浮的泥沙发生淤积，而应当是不冲、不淤的流速。即

$$v_{max} > v > v_{min}$$

式中，v_{max} 为免受冲刷的最大允许流速，简称不冲允许流速；v_{min} 为免淤积的最小允许流速，简称不淤允许流速。

渠道中的不冲允许流速 v_{max} 取决于土质情况或渠道衬砌材料。表 8-4 给出了各种渠道免遭冲刷的最大允许流速，可供设计明渠时使用。

Table 8-4 Non-Scouring Velocities of Open Channels
表 8-4　渠道的不冲允许流速　　　　　　　　　　单位：m/s

Channel types 渠道类型	Types of rock and surface cover 岩石或护面种类	Flow rate in open channel 渠道流量 /(m³/s)		
		<1	1~10	>10
Hard rock and artificial protective channels 坚硬岩石和人工护面渠道	Soft sedimentary rock (marl, shale, and soft conglomerate) 软质水成岩（泥灰岩、页岩、软砾岩）	2.5	3.0	3.5
	Medium hard sedimentary rock (compact conglomerate, porous limestone, layer limestone, dolomitic limestone, and calcareous sandstone) 中等硬质水成岩（致密砾岩、多孔石灰岩、层状石灰岩、白云石灰岩、灰质砂岩）	3.5	4.25	5.0

续表

Channel types 渠道类型	Types of rock and surface cover 岩石或护面种类	Flow rate in open channel 渠道流量 /(m³/s)		
		<1	1~10	>10
Hard rock and artificial protective channels 坚硬岩石和人工护面渠道	Hard sedimentary rock (dolomitic sandstone, hard limestone) 硬质水成岩(白云砂岩、硬质石灰岩)	5.0	6.0	7.0
	Crystalline rock, igneous rock 结晶岩、火成岩	8.0	9.0	10.0
	Single layer stone paving 单层块石铺砌	2.5	3.5	4.0
	Double layer stone paving 双层块石铺砌	3.5	4.5	5.0
	Concrete facing (Sands and gravels in the stream) 混凝土护面(水流中含砂和砾石)	6.0	8.0	10.0

Channel types 渠道类型	Types of rock and surface cover 岩石或护面种类			Note 说明
Soil channels 土质渠道	Homogeneous clayey soil 均质黏质土质	Non-scouring velocity 不冲允许流速 /(m/s)		The values of non-scouring velocity for the hydraulic radius $R=1.0$m are listed in this Table. If hydraulic radius $R\neq 1.0$m, the corresponding non-scouring velocity equals to the product of R^α and the value listed in this Table. For sand, gravel, pebbles, loose loam and clay, $\alpha=1/3\sim 1/4$; while for dense loam and clay, $\alpha=1/4\sim 1/5$ 表中所列为水力半径 $R=1.0$m的情况；$R\neq 1.0$m时,应将表中数据乘以R^α才得相应的不冲允许流速值。对于砂、砾石、卵石、疏松的壤土、黏土,$\alpha=1/4\sim 1/3$;对于密实的壤土、黏土,$\alpha=1/5\sim 1/4$
	Light loam 轻壤土	0.6~0.8		
	Medium loam 中壤土	0.65~0.85		
	Heavy loam 重壤土	0.70~1.0		
	Clay soil 黏土	0.75~0.95		
	Homogeneous cohesionless soil 均质无黏性土质	Soil particle size 粒径 /mm	Non-scouring velocity 不冲允许流速 /(m/s)	
	Very fine sand 极细砂	0.05~0.1	0.35~0.45	
	Fine sand and medium sand 细砂和中砂	0.25~0.5	0.45~0.60	
	Coarse sand 粗砂	0.5~2.0	0.60~0.75	
	Fine gravel 细砾石	2.0~5.0	0.75~0.90	
	Medium gravel 中砾石	5.0~10.0	0.90~1.10	
	Coarse gravel 粗砾石	10.0~20.0	1.10~1.30	
	Fine pebble 小卵石	20.0~40.0	1.30~1.80	
	Medium pebble 中卵石	40.0~60.0	1.80~2.20	

渠道中的不淤允许流速 v_{min} 按以下标准选取：

① 防止植物滋生 $v_{min}=0.6 \text{m/s}$。

② 防止淤泥淤积 $v_{min}=0.2 \text{m/s}$。

③ 防止砂的沉积 $v_{min}=0.4 \text{m/s}$。

8.4 Specific Energy and Critical Depth at a Section

8.4.1 Specific Energy at a Section

As shown in Figure 8-3, for any cross-section in a gradually-varying open channel flow, the total mechanical energy E of unit mass liquid at any point referred to datum 0-0 is $E = z + \frac{p}{\gamma} + \frac{\alpha v^2}{2g}$. If the datum is set at the lowest point of the section, the mechanical energy e of unit mass liquid at any point on water surface referred to the new datum 0_1-0_1 is

$$e = E - z_1 = h + \frac{\alpha v^2}{2g} \tag{8-15}$$

The e is defined as specific energy in hydraulics. Equation (8-15) can be expressed as follows:

$$e = h + \frac{\alpha Q^2}{2gA^2} \tag{8-16}$$

In Equation (8-16), e is a function of depth h. After analysis, we know when $h \to 0$, $A \to 0$ and $e \to \infty$; when $h \to \infty$, $A \to \infty$ and $e \to \infty$. Hence, as shown in Figure 8-4, the depth with respect to e_{min} in the curve of $e = f(h)$ is called critical depth h_K.

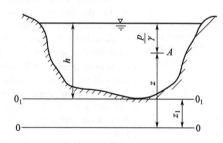

Figure 8-3 Specific Energy at a Cross-Section

图 8-3 断面单位能量

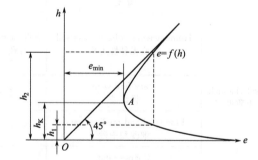

Figure 8-4 Critical Depth

图 8-4 临界水深

8.4 断面单位能量与临界水深

8.4.1 断面单位能量

如图 8-3 所示，在明渠渐变流的任一过水断面中，任意一点的单位重量液体对某一基准面 0—0 的总机械能 E 为 $E = z + \frac{p}{\gamma} + \frac{\alpha v^2}{2g}$。

如果把基准面设在断面最低点，则水面某一点的单位重量液体对新基准面 0_1—0_1 的机械能 e 为

$$e = E - z_1 = h + \frac{\alpha v^2}{2g} \tag{8-15}$$

在水力学中把 e 称为断面单位能量或断面比能，将式(8-15) 表达如下：

$$e = h + \frac{\alpha Q^2}{2gA^2} \tag{8-16}$$

式中，e 为水深 h 的函数。经过一系列分析，当 $h \to 0$ 时，$A \to 0$，此时 $e \to \infty$；当 $h \to \infty$ 时，$A \to \infty$，此时 $e \to \infty$。因此，如图 8-4 所示，曲线 $e = f(h)$ 对应于 $e_{\min}$ 处有一水深，称为临界水深 h_K。

8.4.2 Critical Depth

For a given section shape and flux, critical depth represents the depth of the section at which the specific energy is a minimum.

According to the definition above, critical depth h_K can be calculated. Differentiating e with respect to h at the two sides of Equation (8-16) yields

$$\frac{de}{dh} = 1 - \frac{\alpha Q^2}{gA^3} \frac{dA}{dh}$$

where dA/dh —the change of cross-section area with respect to depth, and it can be substituted by water surface width b approximately, namely, $dA/dh = b$. Thus the equation above changes to

$$\frac{de}{dh} = 1 - \frac{\alpha Q^2 b}{gA^3} \tag{8-17}$$

Let $\dfrac{de}{dh} = 0$, and we can get the universal expression for critical depth h_K when $e = e_{\min}$:

$$\frac{\alpha Q^2}{g} = \frac{A_K^3}{b_K} \tag{8-18}$$

where both A_K and b_K are functions of critical depth, so we can obtain h_K. With h and A^3/b as vertical coordinate and horizontal coordinate, respectively, we give various h values, and calculate corresponding A^3/b values in turn, afterwards plot the curve between h and A^3/b. When A^3/b is equal to $\alpha Q^2/g$, the corresponding depth h just is critical depth h_K.

For a rectangle open channel, we can solve critical depth h_K using the following formula.

$$\frac{\alpha Q^2}{g} = \frac{(bh_K)^3}{b}$$

$$h_K = \sqrt[3]{\frac{\alpha Q^2}{gb^2}} = \sqrt[3]{\frac{\alpha q^2}{g}} \tag{8-19}$$

where $q = Q/b$ is flux for unit width channel.

In a prismatic channel flow, for a given shape, size and flux of cross-section, the channel bed slope is called critical slope i_K when the normal depth h_0 just equals to critical depth h_K.

8.4.2 临界水深

临界水深是指在断面形式和流量给定的条件下，断面单位能量为最小值时的水深。

临界水深 h_K 的计算可根据上述定义得出，对式(8-16)求 e 对 h 的导数得

$$\frac{de}{dh} = 1 - \frac{\alpha Q^2}{g A^3} \frac{dA}{dh}$$

式中，dA/dh 为过流断面面积随水深的变化率，可近似地以水面宽度 b 代替，即 $dA/dh = b$。因而，上式可变为

$$\frac{de}{dh} = 1 - \frac{\alpha Q^2 b}{g A^3} \tag{8-17}$$

令 $\frac{de}{dh} = 0$，得 $e = e_{\min}$ 时临界水深 h_K 的普遍表达式：

$$\frac{\alpha Q^2}{g} = \frac{A_K^3}{b_K} \tag{8-18}$$

式中，A_K、b_K 为临界水深的函数，故可确定 h_K。以 h 为纵坐标，A^3/b 为横坐标，设不同的 h 值，依次求得 A^3/b，绘出曲线，A^3/b 等于 $\alpha Q^2/g$ 时的水深 h 便是 h_K。

对于矩形断面明渠，用下式求临界水深 h_K。

$$\frac{\alpha Q^2}{g} = \frac{(bh_K)^3}{b}$$

$$h_K = \sqrt[3]{\frac{\alpha Q^2}{g b^2}} = \sqrt[3]{\frac{\alpha q^2}{g}} \tag{8-19}$$

式中，$q = \frac{Q}{b}$，称为单宽流量。

在棱柱形渠道中，断面形状、尺寸和流量一定时，水流的正常水深 h_0 恰好等于临界水深 h_K 时，其渠底坡度称为临界坡度 i_K。

8.4.3 Subcritical, Supercritical and Critical Flows

The velocity of flow in an open channel in which the depth just equals to critical depth is known as critical velocity v_K. Such flow is referred to critical flow. If velocities are less than critical, the flow represents subcritical flow. If velocities are greater than critical, the flow represents supercritical flow.

Besides, specific energy also can be used to distinguish whether a flow is subcritical or supercritical:

when $\frac{de}{dh} > 0$, the flow is subcritical;

but when $\frac{de}{dh} < 0$, the flow is supercritical.

The distinguishing of subcritical or supercritical flow is important for analysis and calculation of nonuniform flow in open channel. In addition to critical velocity, critical depth and specific energy, a simpler criterion number, Froude number (Fr), can be used.

In Equation (8-17), the item, $\frac{\alpha Q^2 b}{g A^3}$, as a dimensionless number, is named as Froude number Fr in hydraulics, namely

$$Fr = \frac{\alpha Q^2 b}{gA^3}$$

So Equation (8-17) changes to

$$\frac{de}{dh} = 1 - Fr \tag{8-20}$$

Let $\frac{A}{b} = h_m$, which represents the mean depth at cross-section. Thus Froude number Fr can be written as

$$Fr = \frac{\alpha Q^2}{gA^2 h_m} = \frac{\alpha v^2}{gh_m} = 2\frac{\frac{\alpha v^2}{2g}}{h_m} \tag{8-21}$$

From Equation (8-21), we can know that Froude number Fr denotes the ratio of energy, and it equals to twice of the ratio of kinetic energy of unit weight liquid versus mean potential energy. According to Equation (8-20), we can find:

if $\frac{de}{dh} = 0$, $Fr = 1$, the flow is critical;

if $\frac{de}{dh} < 0$, $Fr > 1$, the flow is supercritical;

and if $\frac{de}{dh} > 0$, $Fr < 1$, the flow is subcritical.

8.4.3 缓流、急流、临界流

明渠水流在临界水深时的流速称为临界流速，以 v_K 表示，这样的明渠水流状态称为临界流。当明渠水流流速小于临界流速时，称为缓流；大于临界流速时，称为急流。

用断面单位能量也可判别缓流和急流：当 $\frac{de}{dh} > 0$ 时，为缓流；当 $\frac{de}{dh} < 0$ 时，为急流。

缓流与急流的判别在明渠非均匀流的分析计算上具有重要意义。除了可用临界流速、临界水深、断面单位能量的变化做判别外，还可用更简单的判别准则——弗劳德数 Fr (Froude number) 来判别。

从式(8-17)中 $\frac{\alpha Q^2 b}{gA^3}$ 项分析，它是一个无量纲组合数，在水力学中称它为弗劳德数，以 Fr 表示，即

$$Fr = \frac{\alpha Q^2 b}{gA^3}$$

所以式(8-17)变为

$$\frac{de}{dh} = 1 - Fr \tag{8-20}$$

令 $\frac{A}{b} = h_m$ 表示过水断面上的平均水深，则弗劳德数 Fr 写为

$$Fr = \frac{\alpha Q^2}{gA^2 h_m} = \frac{\alpha v^2}{gh_m} = 2\frac{\frac{\alpha v^2}{2g}}{h_m} \tag{8-21}$$

由式(8-21)可知，弗劳德数 Fr 代表能量比值，为水流中单位重量液体的动能与其平均势能比值的 2 倍。由式(8-20)可得：

$\dfrac{\mathrm{d}e}{\mathrm{d}h}=0$ 时，$Fr=1$，为临界流；

$\dfrac{\mathrm{d}e}{\mathrm{d}h}<0$ 时，$Fr>1$，为急流；

$\dfrac{\mathrm{d}e}{\mathrm{d}h}>0$ 时，$Fr<1$，为缓流。

Exercises 习题

8-1　A concrete pipe with a diameter of 600mm on a 1/400 slope carries water at one-half full. Using Manning equation with $n=0.013$, find the flow rate Q.

8-1　管径为 600mm 的混凝土管，以坡度为 1/400、半充满状态进行输水，$n=0.013$，应用曼宁公式求流量 Q。

8-2　Water flows uniformly in a 3.0-m-wide rectangular channel at a depth of 500mm. The channel slope is 0.003 and $n=0.012$. Find the flow rate in m^3/s.

8-2　水在深 500mm、宽 3.0m 的矩形渠道中均匀流动，坡度为 0.003，$n=0.012$。求流量，以 m^3/s 计。

8-3　If $b=$ bottom width and $y=$ depth, under which of the following conditions does the best hydraulic rectangular cross-section occur：(a) $y=2b$；(b) $y=b$；(c) $y=b^2$；(d) $y=b/2$.

8-3　b 为宽度，y 为水深，矩形的水力最优断面在下列哪种条件下会发生：(a) $y=2b$；(b) $y=b$；(c) $y=b^2$；(d) $y=b/2$。

8-4　What is the best hydraulic rectangular cross-section defined as：(a) the least expensive cross-section；(b) the section with minimum roughness coefficient；(c) the one that has a minimum perimeter；(d) the section that has a maximum area for a given flow.

8-4　矩形水力最优断面的定义是下列哪个：(a) 造价最少的过流断面；(b) 粗糙系数最小的断面；(c) 湿周最小的断面；(d) 给定流量下面积最大的断面。

8-5　Under which of the following conditions does supercritical flow can never occur：(a) in a steep channel；(b) in a mild channel；(c) in an adverse channel；(d) in a horizontal channel；(e) directly after a hydraulic jump.

8-5　在下列哪些情况下不会发生急流：(a) 陡坡渠道内；(b) 缓坡渠道内；(c) 逆坡渠道内；(d) 水平渠道内；(e) 水跃后。

8-6　Under which of the following conditions does flow at critical depth occur：(a) the velocity is given by $\sqrt{2gy}$；(b) the normal depth and critical depth coincide for a channel；(c) any change in depth requires more specific energy；(d) the specific energy is a maximum for a given discharge；(e) changes in upstream resistance alter downstream conditions.

8-6　在下列哪些情形下会发生临界水深：(a) 速度为 $\sqrt{2gy}$；(b) 正常水深和临界水深一致时；(c) 在水深上的任何变化需要更大的比能；(d) 已知流量时，比能最大；(e) 上游阻力变化引起了下游的变化。

8-7 Which of the following is the critical depth expression in a rectangular channel:
(a) $\sqrt{\dfrac{q}{g}}$; (b) $\sqrt[3]{\dfrac{q^2}{g}}$; (c) $\sqrt{vy}$; (d) $\sqrt{2gy}$.

8-7 下列哪个式子可以表示矩形渠道中的临界水深：(a) $\sqrt{\dfrac{q}{g}}$；(b) $\sqrt[3]{\dfrac{q^2}{g}}$；(c) $\sqrt{vy}$；(d) $\sqrt{2gy}$。

8-8 Calculate the specific energy for the flow expressed by $v=5\text{m/s}$ and $y=2\text{m}$, in m·N/N.

8-8 当 $v=5\text{m/s}$，$y=2\text{m}$ 时，计算该流动的比能，以 m·N/N 计。

Chapter 9 Seepage Flow
第 9 章 渗　　流

9.1 Introduction to Seepage Flow

A seepage flow is the flow which a fluid flows through porous media. Seepage phenomenon is ubiquitous in natural and artificial materials, such as seepage of groundwater, hot water and brine; seepage of oil, natural gas and coal seam gas; seepage of blood micro-circulation and micro-bronchia in animals; transportation of water, gas and sugar in plant; and seepage of a fluid in artificial porous materials (pottery, brick, sand and filler etc.) and so on.

Mechanics of seepage is widely used in applied science and engineering technology, such as mechanics of soil, subterranean hydrology, oil engineering, terrestrial heat engineering, water supply engineering, environmental engineering, chemical engineering and national defense industry and so on.

The characteristics of seepage flow are as follows:

① In porous medium, the surface area of unit volume interstice is very large, and surface effect is distinct, therefore viscosity of a fluid must be considered;

② In subterranean seepage flow, the pressure of a fluid is very large, so the compressibility should be considered usually;

③ In seepage flow, the form of hole in porous medium is complicated, the flow resistance is large, and capillary effect is universal;

④ Sometimes a seepage flow goes with some complicated physical chemistry processes.

9.1 渗流简介

渗流是流体通过多孔介质的流动。渗流现象普遍存在于自然界和人造材料中，如地下水、热水和盐水的渗流；石油、天然气和煤层气的渗流；动物体内的血液微循环和微细支气管的渗流；植物体内水分、气体和糖分的输送；陶瓷、砖石、砂、填料等人造多孔材料中的渗流等。

渗流力学在很多应用科学和工程技术领域有着广泛的应用，如土壤力学、地下水文学、石油工程、地热工程、给水工程、环境工程、化工和国防工业等。

渗流主要有以下特点：

① 多孔介质单位体积空隙的表面积较大，表面作用明显，因此必须考虑黏性作用。

② 地下渗流中往往压力较大，因而通常要考虑流体的压缩性。

③ 孔道形状复杂，阻力大，毛细作用较普遍。

④ 往往伴随有复杂的物理化学过程。

9.2 Basic Law of Seepage Flow

9.2.1 Seepage Model

Experimental equipment of seepage flow is shown in Figure 9-1. The cylinder is full of sands. The interface area is A and sand layer height is l. In the bottom of sand layer, a thin net is set to support these sands. Water from steady pressure tank flows through valve A to cylinder C. Neglecting kinetic energy, the difference of head H_1-H_2 is head loss of the seepage flow.

9.2 渗流基本定律

9.2.1 渗流模型

图 9-1 为渗流实验装置，圆筒内充填砂粒，截面积为 A，砂层厚为 l，砂底部有细网支撑。水由稳压箱经阀 A 进入圆筒 C，忽略动能，测压管水头差 H_1-H_2 为渗流的水头损失。

9.2.2 Darcy's Law for Seepage Flow

Henri Darcy, a French engineer, carried out many experiments using sandy soil in 1852-1855, and achieved a linear law of seepage flow, namely, Darcy's law.

$$Q = kA \frac{h_w}{l} \tag{9-1}$$

$$v = \frac{Q}{A} = k \frac{h_w}{l} = kJ \tag{9-2}$$

where k — coefficient of seepage flow, m/s;

J — hydraulic slope.

Darcy's law can be applied only for linear seepage where the range of Re is 1~10.

9.2.2 达西渗流定律

法国工程师达西（Henri Darcy）在 1852—1855 年利用沙质土壤进行了大量实验，得到线性渗流定律，即达西定律：

$$Q = kA \frac{h_w}{l} \tag{9-1}$$

$$v = \frac{Q}{A} = k \frac{h_w}{l} = kJ \tag{9-2}$$

式中，k 为渗流系数，m/s；J 为水力坡度。

达西定律适用于线性渗流，Re 为 1~10。

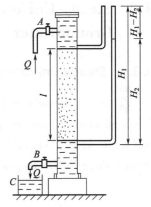

Figure 9-1 Experimental Equipment of Seepage Flow

图 9-1 渗流实验装置

9.2.3 Coefficient of Seepage Flow

Determination methods for coefficient of seepage flow (Table 9-1) are as follows:

① Empirical formula method: estimate coefficient of seepage flow using empirical formula according to soil diameter, shape, structure, porosity, and temperature etc.;

② Experimental method: make experiments using the experimental equipment as shown in Figure 9-1 and calculate coefficient of seepage flow using Equation (9-1);

③ In-site method: make experiments by pumping water from well, and calculate k according to the formula of well.

Table 9-1 Seepage Flow Coefficient of Water in Soil

表 9-1 水在土壤中的渗流系数

Soil species （土壤种类）	Coefficient of seepage flow k/(cm/s) 渗流系数 k/(cm/s)	Soil species （土壤种类）	Coefficient of seepage flow k/(cm/s) 渗流系数 k/(cm/s)
Clay（黏土）	6×10^{-6}	Scree（卵石）	$1 \times 10^{-1} \sim 6 \times 10^{-1}$
Sub-clay（亚黏土）	$6 \times 10^{-6} \sim 1 \times 10^{-4}$	Fine sand（细砂）	$1 \times 10^{-3} \sim 6 \times 10^{-3}$
Loess（黄土）	$3 \times 10^{-4} \sim 6 \times 10^{-4}$	Coarse sand（粗砂）	$2 \times 10^{-3} \sim 6 \times 10^{-2}$

9.2.3 渗流系数

渗流系数（表 9-1）的测定方法如下。

① 经验公式法：根据土壤粒径、形状、结构、孔隙率、温度等参数所组成的经验公式估算。

② 实验室方法：在实验室利用类似图 9-1 的装置及式(9-1) 计算。

③ 现场方法：在现场利用钻井或原有井做抽水或灌水实验，根据水井的计算公式求 k 值。

9.3 Steady Uniform and Steady Gradually-Varying Seepage Flows of Groundwater

9.3.1 Steady Uniform Seepage Flow

There is a steady uniform flow with a bottom slope i (> 0). Due to steady uniform flow, all stream lines are parallel to each other and parallel to impermeable layer. Besides, hydraulic slope J (namely piezometric slope) is equal to bottom slope i. Therefore, the hydraulic slope in the seepage area is equal, i.e.,

$$J = -\frac{dH}{ds} = i \tag{9-3}$$

According to Darcy's law, velocity of seepage flow u is equal in the whole seepage area, namely

$$u = kJ = ki = \text{constant} \tag{9-4}$$

So, the mean velocity v of steady uniform flow is

$$v = u = ki \tag{9-5}$$

9.3 地下水的恒定均匀渗流和恒定渐变渗流

9.3.1 恒定均匀渗流

设一恒定均匀渗流，渠底坡度 $i > 0$，因为是均匀渗流，所有流线都是相互平行的直线，且平行于不透水层，另外，水力坡度 J（即测压管坡度）与渠底坡度 i 相等。因此，在整个

渗流区域内水力坡度相等，即

$$J = -\frac{dH}{ds} = i \tag{9-3}$$

由达西定律可知，渗流流速 u 在整个渗流区域内都相等，即

$$u = kJ = ki = 常数 \tag{9-4}$$

所以恒定均匀渗流的断面平均速度 v 为

$$v = u = ki \tag{9-5}$$

9.3.2 Steady Gradually-Varying Seepage Flow

A gradually-varying seepage flow with a bottom slope i (>0) is shown in Figure 9-2. Due to gradually-varying seepage flow, all stream lines are almost parallel straight lines, so the flow cross-section can be regarded as a plane. At the same flow cross-section, the dynamic pressure distribution accords with the rule of static pressure distribution, and the piezometric head at any point is equal. The velocity of seepage flow is far less and the velocity head can be neglected, so the total head at the same flow cross-section is also equal. Therefore, the head losses in a same stream line between section 1-1 and 2-2 are also equal, and are denoted as head difference dH. Moreover, the curvature of stream line in a gradually-varying flow is less, so the length of a random stream line between two sections may be thought to be an equal value, and is denoted as bottom distance ds. Thus, the hydraulic slope J at any point at a same flow cross-section is equal, and the velocity of seepage flow u is

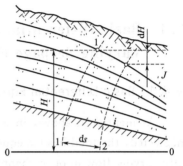

Figure 9-2　Piezometric Slope
图 9-2　测压管坡度

$$u = kJ = -k\frac{dH}{ds} = \text{constant} \tag{9-6}$$

The formula above shows the diagram of velocity distribution is a rectangle, so the mean velocity v is equal to the velocity of seepage flow u at any point at the same section, namely

$$v = u = -k\frac{dH}{ds} \tag{9-7}$$

The formula above is called Dupuit's formula. Dupuit's formula is not suitable for rapidly-varying seepage flow in which the curvature of stream line is very big.

9.3.2 恒定渐变渗流

设一恒定渐变渗流，渠底坡度 $i>0$，如图 9-2 所示。因为是渐变渗流，所有流线近似于平行直线，过流断面可视为平面；同一过流断面上的动水压强分布符合静水压强分布规律，各点测压管水头相等。因为渗流流速较小，流速水头可忽略不计，所以同一过流断面上的总水头也相等。因而，断面 1—1 和 2—2 任一流线上的水头损失也相等，以水头差 dH 表示。另外，因渐变渗流的流线曲率较小，两断面间任一流线的长度可近似地认为相等，并以渠底距离 ds 表示，所以同一过流断面上各点水力坡度 J 也相等，因而同一过流断面上各点的渗流速度 u 为

$$u = kJ = -k\frac{dH}{ds} = 常数 \tag{9-6}$$

上式表明渐变渗流流速分布图为矩形，所以断面平均流速 v 与同一断面上各点流速 u 相等，即

$$v = u = -k\frac{dH}{ds} \tag{9-7}$$

上式称为裘布依（J. Dupuit）公式。裘布依公式不适用于流线曲率很大的急变渗流。

9.4 Basic Differential Equation and Seepage Lines of Gradually-Varying Seepage Flow

9.4.1 Basic Differential Equation of Gradually-Varying Seepage Flow

As mentioned above, in a gradually-varying seepage flow, if the less velocity head is neglected, the total head is equal to the piezometric head, $H = z + h$. Compared with gradually-varying flow in open channel, it is different that in a gradually-varying seepage flow the piezometric slope (namely the slope of line of seepage) can be used to substitute hydraulic slope J. The relation between hydraulic slope J and bottom slope i is shown in Figure 9-2 and it is

$$J = -\frac{dH}{ds} = -\left(\frac{dz}{ds} + \frac{dh}{ds}\right) = i - \frac{dh}{ds}$$

where z — the height of bottom on top of the datum plane 0-0;
 h — water height.

Substituting the formula above into Equation (9-7) yields

$$v = kJ = -k\frac{dH}{ds} = k\left(i - \frac{dh}{ds}\right) \tag{9-8}$$

The flux of gradually-varying seepage flow $Q = vA$, namely

$$Q = kA\left(i - \frac{dh}{ds}\right) \tag{9-9}$$

The formula above is called basic differential equation of gradually-varying seepage flow. It also can be used for horizontal slope and reverse slope of gradually-varying seepage flow of groundwater. For horizontal slope, $i = 0$; for rising slope, $i < 0$.

9.4 渐变渗流基本微分方程和浸润曲线

9.4.1 渐变渗流的基本微分方程

如前所述，在渐变渗流中，如忽略数值较小的流速水头，则总水头与测压管水头相等，即 $H = z + h$。与地面明渠渐变流相比，其不同之处在于可用测压管坡度（即浸润曲线坡度）代替水力坡度 J，而水力坡度与渠底坡度 i 的关系如图 9-2 所示，为

$$J = -\frac{dH}{ds} = -\left(\frac{dz}{ds} + \frac{dh}{ds}\right) = i - \frac{dh}{ds}$$

式中，z 为渠底在 0—0 基准面上的高度；h 为水深。

将上式代入裘布依公式 [式(9-7)]，可得

$$v = kJ = -k\frac{dH}{ds} = k\left(i - \frac{dh}{ds}\right) \tag{9-8}$$

渐变渗流的流量 $Q = vA$，即

$$Q = kA\left(i - \frac{dh}{ds}\right) \tag{9-9}$$

上式即为渐变渗流的基本微分方程。它对于平坡、逆坡地下渐变渗流都适用：平坡时，$i=0$；逆坡时，$i<0$。

9.4.2　Seepage Lines of Gradually-Varying Seepage Flow

For Darcy's law, $Re = 1 \sim 10$, so the velocity is low. Compared with depth, the velocity head can be neglected, therefore the specific energy at a section is equal to depth h. The critical depth loses its meaning, so type of seepage line is less than that of flow profile in an open channel, and there are only four seepage lines during three bottom slopes.

(1) Seepage line of falling slope ($i > 0$)　　As shown in Figure 9-3, in falling slope ($i > 0$) seepage flow, the uniform seepage flow is possible to occur, so that the flux of gradually-varying seepage flow can be replaced by that of the corresponding uniform seepage flow. Substituting $Q = kiA_0$ into basic differential equation of gradually-varying seepage flow [Equation (9-9)] yields

$$Q = kiA_0 = bh_0 ki = bhk\left(i - \frac{dh}{ds}\right)$$

We obtain

$$\frac{dh}{ds} = i\left(1 - \frac{1}{\eta}\right) \tag{9-10}$$

where h_0 is normal depth of falling slope seepage flow; b is width of falling slope seepage flow; h is actual depth of falling slope seepage flow; and $\eta = \dfrac{h}{h_0}$.

As shown in Figure 9-3, there are two sections a and b in falling slope seepage flow.

Due to section a upside of normal depth N-N, $h > h_0$, namely, $\eta > 1$. According to Equation (9-10), $\dfrac{dh}{ds} > 0$, so the depth of seepage line increases with the stream and the seepage line is a back-water curve.

Due to section b downside of normal depth N-N, $h < h_0$, namely, $\eta < 1$. According to Equation (9-10), $\dfrac{dh}{ds} < 0$, so the depth of seepage line decreases with the stream and the seepage line is a dropdown curve.

The depths at section 1-1 and 2-2 are h_1 and h_2, respectively; the distances departed from original section are s_1 and s_2; and the distance between section 1-1 and 2-2 is $l = s_2 - s_1$. Equation (9-10) can be written as follows:

$$\frac{i\,ds}{h_0} = d\eta + \frac{d\eta}{\eta + 1}$$

Integrating the formula above between section 1-1 and 2-2 yields

$$\frac{il}{h_0} = \eta_2 - \eta_1 + 2.3\lg\frac{\eta_2 - 1}{\eta_1 - 1} \tag{9-11}$$

Equation (9-11) is the seepage line equation of falling slope seepage flow.

(2) Seepage line of rising slope ($i < 0$) As shown in Figure 9-4, the seepage line equation of rising slope is

$$\frac{i'l}{h'_0} = \zeta_1 - \zeta_2 + 2.3\lg\frac{1+\zeta_2}{1+\zeta_1} \tag{9-12}$$

where $i' = -i$; h'_0 is normal depth with the bottom slope i'; and $\zeta = \dfrac{h}{h'_0}$.

(3) Seepage line of horizontal slope ($i = 0$) As shown in Figure 9-5, the seepage line equation of horizontal slope is

$$\frac{2q}{k}l = h_1^2 - h_2^2 \tag{9-13}$$

where q is seepage flux per unit width, namely, $q = \dfrac{Q}{b}$.

9.4.2 渐变渗流的浸润曲线

对于达西渗流定律，由于 $Re = 1 \sim 10$，流速是很小的，流速水头和水深相比可以忽略不计，所以断面单位能量实际上就等于水深 h，临界水深失去了意义，故浸润曲线形式比明渠水面曲线少，在三种坡度情况下只有四条浸润曲线。

（1）顺坡（$i > 0$）浸润曲线 如图 9-3 所示，在 $i > 0$ 的顺坡渗流中，有可能发生均匀渗流，所以渐变渗流的流量可用相应的均匀渗流的流量代替，将 $Q = kiA_0$ 代入渐变渗流基本微分方程［式(9-9)］，则

$$Q = kiA_0 = bh_0 ki = bhk\left(i - \frac{\mathrm{d}h}{\mathrm{d}s}\right)$$

得

$$\frac{\mathrm{d}h}{\mathrm{d}s} = i\left(1 - \frac{1}{\eta}\right) \tag{9-10}$$

式中，h_0 为顺坡渗流正常水深；b 为顺坡渗流宽度；h 为顺坡渗流实际水深；$\eta = \dfrac{h}{h_0}$。

在顺坡渗流中分为 a、b 两区，见图 9-3。

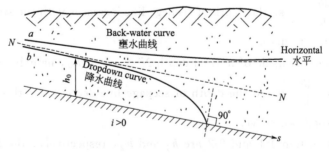

Figure 9-3 Falling Slope Seepage Flow
图 9-3 顺坡渗流

在正常水深 $N—N$ 之上 a 区的浸润曲线，$h > h_0$，即 $\eta > 1$，由式(9-10) 可见 $\dfrac{\mathrm{d}h}{\mathrm{d}s} > 0$，浸润曲线的水深是沿流向增加的，为壅水曲线。

在正常水深 $N—N$ 以下的 b 区的浸润曲线，$h<h_0$，即 $\eta<1$，由式(9-10) 可见 $\dfrac{dh}{ds}<0$，浸润曲线的水深是沿流向减小的，为降水曲线。

对于断面 1—1 和断面 2—2，水深为 h_1 及 h_2，距起始断面沿坡底方向距离为 s_1 及 s_2，两断面相距 $l=s_2-s_1$，由式(9-10) 得

$$\frac{i\,ds}{h_0}=d\eta+\frac{d\eta}{\eta+1}$$

在断面 1—1 及 2—2 间积分，得

$$\frac{il}{h_0}=\eta_2-\eta_1+2.3\lg\frac{\eta_2-1}{\eta_1-1} \qquad (9\text{-}11)$$

此即顺坡平面渗流浸润曲线方程。

(2) 逆坡（$i<0$）浸润曲线　如图 9-4 所示，浸润曲线方程为

$$\frac{i'l}{h'_0}=\zeta_1-\zeta_2+2.3\lg\frac{1+\zeta_2}{1+\zeta_1} \qquad (9\text{-}12)$$

式中，$i'=-i$；h'_0 为 i' 坡度上的正常水深；$\zeta=\dfrac{h}{h'_0}$。

(3) 平坡（$i=0$）浸润曲线　如图 9-5 所示，该浸润曲线方程为

$$\frac{2q}{k}l=h_1^2-h_2^2 \qquad (9\text{-}13)$$

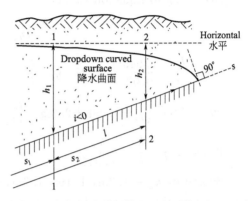

Figure 9-4　Rising Slope Seepage Flow

图 9-4　逆坡渗流

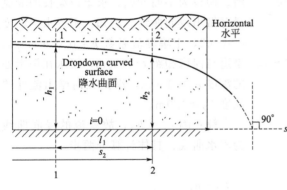

Figure 9-5　Horizontal Slope Seepage Flow

图 9-5　平坡渗流

式中，$q=\dfrac{Q}{b}$，即单宽渗流流量。

【Sample Problem 9-1】 As shown in Figure 9-6, an open channel is located upside of a river, and water in open channel seeps through a side of bank down to the river. Supposing that the seepage flow is a plane flow, try to calculate the seepage flux per unit length and plot the seepage line. Given that the impermeable layer slope $i=0.02$, coefficient of seepage flow in soil $k=0.005$ cm/s, the distance between channel and riverbed $l=180$ m, the depth on the channel bank $h_1=1.0$ m, and the depth of seepage flow at the stream exit $h_2=1.9$ m.

Solution: $h_1<h_2$, so the seepage line is a back-water curve. The calculation includes two steps.

① Calculation of seepage flux. According to Equation (9-11), we can obtain

$$il-h_2+h_1=2.3h_0\lg\frac{h_2-h_0}{h_1-h_0}$$

Namely $$h_0 \lg \frac{1.9 - h_0}{1.0 - h_0} = \frac{1}{2.3}(0.02 \times 180 - 1.9 + 1.0) = 1.174$$

By trial calculation, we obtain $h_0 = 0.945$m, therefore
$$q = h_0 v_0 = kih_0 = 0.005 \times 0.02 \times 0.945 \times 100 = 0.00945 \text{cm}^3/(\text{s} \cdot \text{cm})$$

② Calculation of seepage line. From the channel bank to the river bank, several h_2 ($1.0\text{m} < h_2 < 1.9\text{m}$) are given in turn and calculate the distances l between the location with the depth of h_2 and the channel bank.

According to Equation (9-11)
$$l = \frac{h_0}{i}\left(\eta_2 - \eta_1 + 2.3\lg\frac{\eta_2 - 1}{\eta_1 - 1}\right)$$

where $\dfrac{h_0}{i} = \dfrac{0.945}{0.02} = 47.25$, $\eta_1 = \dfrac{h_1}{h_0} = \dfrac{1}{0.945} = 1.058$.

So $l = 47.25\left(\eta_2 - 1.058 + 2.3\lg\dfrac{\eta_2 - 1}{1.058 - 1}\right)$.

Besides, $\eta_2 = \dfrac{h_2}{h_0} = \dfrac{h_2}{0.945}$.

Substitute $h_2 = 1.2$m, 1.4m, 1.7m, and 1.9m into the formula above respectively, we can calculate the corresponding l, and the corresponding l are 82.6m, 120m, 159m and 180m, respectively. The results are plotted in Figure 9-6.

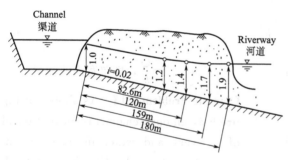

Figure 9-6 Figure for Sample Problem 9-1
图 9-6 例题 9-1 图

【例题 9-1】 如图 9-6 所示，一渠道位于河道上方，渠水沿岸的一侧下渗入河。假设为平面问题，求单位渠长的渗流量并做出浸润曲线。已知：不透水层坡度 $i = 0.02$，土壤渗流系数 $k = 0.005$ cm/s，渠道与河床相距 $l = 180$m，渠水在渠岸处的深度 $h_1 = 1.0$m，渗流在河岸出流处的深度 $h_2 = 1.9$m。

解：因 $h_1 < h_2$，故渗流的浸润曲线为壅水曲线，具体计算分两步。

① 计算渠岸渗流量。由式(9-11) 得
$$il - h_2 + h_1 = 2.3 h_0 \lg \frac{h_2 - h_0}{h_1 - h_0}$$

即 $$h_0 \lg \frac{1.9 - h_0}{1.0 - h_0} = \frac{1}{2.3}(0.02 \times 180 - 1.9 + 1.0) = 1.174$$

试算得 $h_0 = 0.945$m，从而
$$q = h_0 v_0 = Kih_0 = 0.005 \times 0.02 \times 0.945 \times 100$$
$$= 0.00945 \text{cm}^3/(\text{s} \cdot \text{cm})$$

② 计算浸润曲线。从渠岸往下游算至河岸为止，上游水深 $h_1 = 1.0$m，依次给出 1.0m $< h_2 < 1.9$m 的几种渐增值，分别算出各个 h_2 处距上游的距离 l。由式(9-11) 得
$$l = \frac{h_0}{i}\left(\eta_2 - \eta_1 + 2.3\lg\frac{\eta_2 - 1}{\eta_1 - 1}\right)$$

式中，$\dfrac{h_0}{i}=\dfrac{0.945}{0.02}=47.25$；$\eta_1=\dfrac{h_1}{h_0}=\dfrac{1}{0.945}=1.058$。

则 $l=47.25\left(\eta_2-1.058+2.3\lg\dfrac{\eta_2-1}{1.058-1}\right)$。

又 $\eta_2=\dfrac{h_2}{h_0}=\dfrac{h_2}{0.945}$，并将 $h_2=1.2\text{m}$、1.4m、1.7m、1.9m 代入，便可求得相应的 l 为 82.6m、120m、159m、180m，其结果绘于图 9-6 中。

9.5 Catchment Passage and Well

9.5.1 Catchment Passage

As shown in Figure 9-7, there is a catchment passage of which the cross-section is rectangle, and the bottom is an impermeable layer with a slope of $i=0$.

According to Equation (9-9), we obtain

$$Q=kA\left(0-\dfrac{\mathrm{d}h}{\mathrm{d}s}\right)=-bhk\dfrac{\mathrm{d}h}{\mathrm{d}s}$$

Integrating the formula above yields

$$z^2-h^2=\dfrac{2Q}{kb}x \tag{9-14}$$

The formula above is the seepage line equation of catchment passage.

When $x\geqslant L$, the groundwater is not affected, so L is called influencing range of catchment passage.

Substituting $x=L$ and $z=H$ into Equation (9-14) yields the seepage flux through a single side.

$$q=\dfrac{Q}{b}=\dfrac{k(H^2-h^2)}{2L} \tag{9-15}$$

9.5 集水廊道和井

9.5.1 集水廊道

如图 9-7 所示，某集水廊道，横断面为矩形，底为不透水层，底坡 $i=0$。

由式(9-9)得

$$Q=kA\left(0-\dfrac{\mathrm{d}h}{\mathrm{d}s}\right)=-bhk\dfrac{\mathrm{d}h}{\mathrm{d}s}$$

积分得

$$z^2-h^2=\dfrac{2Q}{kb}x \tag{9-14}$$

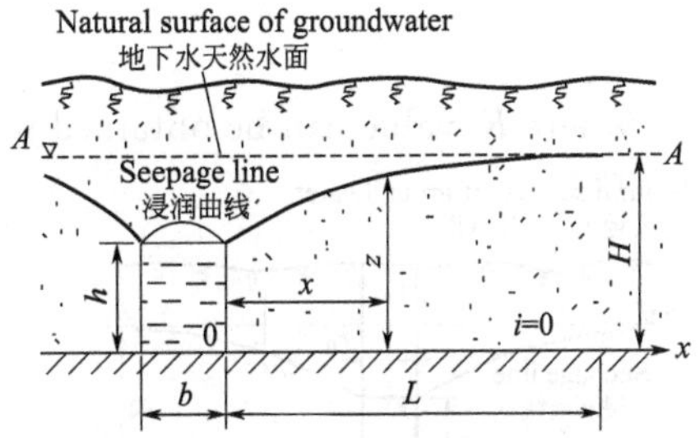

上式即为集水廊道的浸润曲线方程。

当 $x\geqslant L$ 时，地下水不受影响，称 L 为集水廊道的影响范围。

将 $x=L$、$z=H$ 代入式(9-14)，可得单侧渗流量为

$$q=\dfrac{Q}{b}=\dfrac{k(H^2-h^2)}{2L} \tag{9-15}$$

Figure 9-7 Catchment Passage

图 9-7 集水廊道

9.5.2 Underflow Well

In spite of purposes of well, every well used to draw nonpressure groundwater with a free surface on the top of impermeable layer is called underflow well, while the well used to draw pressure groundwater between two impermeable layers is called artesian well. If the bottom of the well is on the impermeable layer, the well is called completely penetrating well, or else, the well is called incompletely penetrating well.

A completely penetrating underflow well on the top of an impermeable layer is shown in Figure 9-8. For a completely penetrating underflow well, the Dupuit's equation can be used to analyze and calculate.

Given that when the radius apart from center axis is r, the vertical coordinate of a seepage line at a cross-section is z. When the radius is $r+dr$, the vertical coordinate is $z+dz$ and dz is positive. According to characteristics of gradually-varying seepage flow, we can know the hydraulic slope J of any point at this section is equal, namely

$$J = \frac{dz}{dr}$$

The mean velocity $v = kJ$, so $v = k\frac{dz}{dr}$.

And the seepage flux Q through the section is $Q = kA\frac{dz}{dr} = 2\pi rzk\frac{dz}{dr}$.

Integrating the formula above yields

$$z^2 - h^2 = \frac{Q}{\pi k}\ln\frac{r}{r_0} \tag{9-16}$$

When $r = R$, $z = H$. R is called the influencing radius of well, and the groundwater outside of R is not affected by pumping.

Substituting $r = R$ and $z = H$ into Equation (9-16) yields

$$Q = \frac{k\pi(H^2 - h^2)}{\ln\frac{R}{r_0}} \tag{9-17}$$

where R value can be obtained using experimental method. If there are no experimental data, during primary calculation, empirical formula can be used to estimate R, namely

$$R = 3000s\sqrt{k} \tag{9-18}$$

where $s = H - h_0$, and s is surface drop depth, m; k is coefficient of seepage flow in soil, m/s.

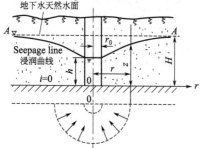

Figure 9-8 Completely Penetrating Underflow Well

图 9-8 完全潜水井

9.5.2 潜水井

不论井的用途如何，凡是汲取位于不透水层上部且具有自由浸润面的无压地下水的井均称为潜水井，而汲取两不透水层之间的有压地下水的井称为自流井。井底直达不透水层的称为完全井，否则称为不完全井。

设位于不透水层上的完全潜水井如图 9-8 所示。

对于完全潜水井，可采用裘布依公式进行分析和计算。

设距井中心轴的半径为 r 处的某过流断面的浸润曲线纵坐标为 z，当半径为 $r+dr$ 时，纵坐标为 $z+dz$，且 dz 为正值。由渐变渗流特性可知该断面各点的水力坡度 J 均相等，即

$$J = \frac{dz}{dr}$$

因断面平均流速 $v = kJ$，所以 $v = k\dfrac{dz}{dr}$。

通过该断面的渗流量为

$$Q = kA\frac{dz}{dr} = 2\pi rzk\frac{dz}{dr}$$

上式积分，可得

$$z^2 - h^2 = \frac{Q}{\pi k}\ln\frac{r}{r_0} \tag{9-16}$$

当 $r=R$ 时，$z=H$，称此 R 为井的影响半径，距离 R 以外的地下水将不受该井抽水的影响。将 $r=R$、$z=H$ 代入式(9-16) 可得

$$Q = \frac{k\pi(H^2 - h^2)}{\ln\dfrac{R}{r_0}} \tag{9-17}$$

式中的 R 需用实验方法求得。无实验资料的情况下，在初步计算时，可用经验公式估算，即

$$R = 3000s\sqrt{k} \tag{9-18}$$

式中，$s = H - h_0$，s 为抽水后井中水面降落深度，m；k 为土壤渗流系数，m/s。

9.5.3 Complete Artesian Well

A complete artesian well is shown in Figure 9-9. Choose a cross-section of seepage flow with a distance of r from the center axis of the well. The area of the cross-section $A = 2\pi rt$ and it is independent of piezometric head z. According to characteristics of gradually-varying seepage flow, we can know the hydraulic slope J of any point at this section is equal, namely, $J = \dfrac{dz}{dr}$, so the flux through the section is

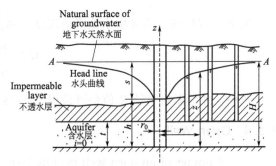

Figure 9-9 Complete Artesian Well
图 9-9 完全自流井

$$Q = 2\pi rtk\frac{dz}{dr}$$

Separating variables in the formula above and integrating yields

$$z - h = \frac{Q}{2\pi kt}\ln\frac{r}{r_0} \tag{9-19}$$

The formula above is head line equation of complete artesian well.

Substitute $z = H$ and $r = R$ into the formula above, and we can obtain the seepage flux of well, namely

$$Q = \frac{2\pi(H-h)\,kt}{\ln\dfrac{R}{r_0}} \tag{9-20}$$

Because the surface drop depth $s = H - h$, the formula above also can be written as the following equation.

$$Q = \frac{2\pi skt}{\ln\dfrac{R}{r_0}} \tag{9-21}$$

In this formula, the influencing radius R also can be estimated using Equation (9-18).

9.5.3 完全自流井

设一完全自流井如图 9-9 所示。取距井中心轴为 r 的渗流过流断面，该断面面积 $A = 2\pi rt$，它与测压管水头 z 无关。由渐变渗流特性可知该断面各点的水力坡度 J 均相等，$J = \dfrac{\mathrm{d}z}{\mathrm{d}r}$，通过该断面的流量为

$$Q = 2\pi rtk\frac{\mathrm{d}z}{\mathrm{d}r}$$

上式分离变量并积分，可得

$$z - h = \frac{Q}{2\pi kt}\ln\frac{r}{r_0} \tag{9-19}$$

上式为完全自流井的水头曲线方程式。

将 $z = H$、$r = R$ 代入上式，就可求出井的渗流量 Q，即

$$Q = \frac{2\pi(H-h)\,kt}{\ln\dfrac{R}{r_0}} \tag{9-20}$$

由于 $s = H - h$，为井中水面降落深度，则上式也可写成

$$Q = \frac{2\pi skt}{\ln\dfrac{R}{r_0}} \tag{9-21}$$

式中，井的影响半径 R 值仍可采用式(9-18)估算。

9.5.4 Open Well

The diameter of an open well is quite large. Usually the diameter is more than 2m, and sometimes the diameter even exceeds 10m.

If the bottom of an open well reaches the impermeable layer, and inflow of the well is not from the bottom, just through the well wall, this open well is called complete open well, and the calculation of its flux is the same as that of deep well (completely penetrating well) to draw groundwater at deep level. The flux calculations of pumping nonpressure groundwater and groundwater on confined aquifer can use Equation (9-17) and Equation (9-21), respectively.

As shown in Figure 9-10, the bottom of an open well does not reach impermeable layer, while is located on the shallow aquifer. This open well is called incomplete open well. When the aquifer is very thick, water flows into open well through the half spherical bottom and seepage stream flows along the radius of the half spherical surface. So the cross-section is a homocentric half spherical surface with the bottom. Its area $A = 2\pi r^2$, and hydraulic slope $J = \dfrac{dz}{dr}$. So the flux of the open well Q is

$$Q = 2\pi r^2 k \dfrac{dz}{dr}$$

For two boundary conditions ($r = r_0$, $z = H - s$; $r = R$, $z = H$), separating variables in the formula above and integrating yields

$$Q \int_{r_0}^{R} \dfrac{dr}{r^2} = 2\pi k \int_{H-s}^{H} dz$$

We can obtain

$$Q = \dfrac{2\pi k s}{\dfrac{1}{r_0} - \dfrac{1}{R}}$$

The influencing radius R is far larger than the well radius r_0, thus the formula above can be simplified to the following equation.

$$Q = 2\pi k r_0 s \tag{9-22}$$

The formula above is the equation to calculate the flux of incomplete open well with a half spherical bottom. In this formula, s is the surface drop depth in open well with the condition of steady pumping water.

As shown in Figure 9-11, for an incomplete open well with a flat bottom, it is believed that the cross-section is a half ellipsoid and the stream line is a hyperbola. The calculating equation for flux of an incomplete open well with a flat bottom is

$$Q = 4 k r_0 s \tag{9-23}$$

During construction process, when footing groove is used to drain groundwater, the relation between discharge and water level drop can be calculated as an open well.

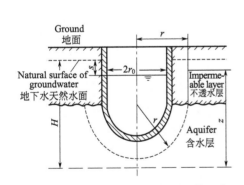

Figure 9-10 Incomplete Open Well with a Half Spherical Bottom
图 9-10 井底为半球形的不完全大口井

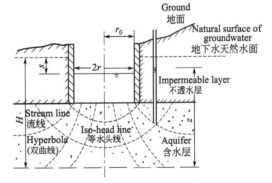

Figure 9-11 Incomplete Open Well with a Flat Bottom
图 9-11 平底不完全大口井

9.5.4 大口井

大口井的直径一般较大，通常在 2m 以上，有时甚至超过 10m。

若大口井的井底直达不透水层，底部不能进水，只是侧壁进水，则此种大口井为完全大口井，其流量计算与汲取深层地下水的管井（完全井）相同。汲取无压地下水的可用式(9-

17) 计算流量, 汲取承压含水层地下水的可用式(9-21) 计算流量。

若大口井的井底未达不透水层, 而是在较浅的含水层中, 如图 9-10 所示, 此种大口井为不完全大口井。若含水层的厚度很大, 水由大口井半球形底部流入, 渗流流线沿半球面的半径方向, 过流断面为与井底半球同心的半球面, 面积 $A = 2\pi r^2$, 水力坡度 $J = \dfrac{dz}{dr}$, 所以井的出水量 Q 为

$$Q = 2\pi r^2 k \frac{dz}{dr}$$

上式分离变量后积分, 当 $r = r_0$ 时, $z = H - s$, 当 $r = R$ 时, $z = H$, 则

$$Q \int_{r_0}^{R} \frac{dr}{r^2} = 2\pi k \int_{H-s}^{H} dz$$

可得

$$Q = \frac{2\pi k s}{\dfrac{1}{r_0} - \dfrac{1}{R}}$$

影响半径远大于井的半径 r_0 值, 因此上式可简化为

$$Q = 2\pi k r_0 s \tag{9-22}$$

式(9-22) 为底部半球进水的不完全大口井的流量计算式, 式中 s 为大口井抽水稳定后井中水面的降落深度。

对于平底不完全大口井, 有学者认为过水断面是半椭球面, 流线是双曲线, 如图 9-11 所示。平底不完全大口井的出水量 Q 的计算公式为

$$Q = 4 k r_0 s \tag{9-23}$$

在施工过程中, 用基坑排出地下水时, 出水量与水位降落的关系可按大口井来计算。

Comparison between English and Chinese Terms
英汉术语对照

A

absolute pressure	绝对压强	angular deformation	角变形
absolute roughness	绝对粗糙度	angular velocity	角转速
absolute temperature	绝对温度	anisotropy	各向异性
absolute velocity	绝对速度	aquifer	含水层
acceleration	加速度	artesian well	自流井
acting force	作用力	artificial channel	人工渠道
adhesion	附着力	artificial roughness	人工粗糙度
analysis of flow profile	水面曲线分析	atmospheric pressure	大气压强
analytical method	分析方法		

B

back flow	回流	bottom slope	底坡
back-water curve	壅水曲线	boundary condition	边界条件
Bernoulli's equation	伯努利方程	branching pipes	枝状管网
body force	质量力	buoyancy	浮力

C

capillary phenomena	毛细现象	concentration gradient	浓度梯度
cavitation	空化	confined aquifer	承压含水层
cavitation damage	气蚀	conservation of mass	质量守恒
center of gravity	重心	conservation of mechanical energy	机械能守恒
channel of compound cross-section	复式断面渠道	continuity equation	连续性方程
characteristic numbers	特征数	continuous medium hypothesis	连续介质假设
Chézy formula	谢才公式	continuous medium	连续介质
coefficient of permeability	渗透系数	contractional depth	收缩断面水深
coefficient of roughness	粗糙系数	control depth	控制水深
cohesion	内聚力	control section	控制断面
completely penetrating well	完全井	control surface	控制面
compressible fluid	可压缩流体	control volume	控制体
compressible gas	可压缩气体	convection acceleration	迁移加速度
computation of flow profile	水面曲线计算	critical depth	临界水深
computational fluid mechanics	计算流体力学	critical flow	临界流
concentration	浓度	critical Reynolds number	临界雷诺数
concentration field	浓度场	critical slope	临界坡
concentration fluctuation	脉动浓度	critical velocity	临界流速

D

D'Alembert paradox	达朗贝尔佯谬	discharge	排放量
Darcy's law	达西定律	discharge per unit width	单宽流量
density	密度	discharge scale	流量比尺
diagram of pressure distribution	压强分布图	distribution of velocity	流速分布
differential pressure	压差	drag coefficient	阻力系数
diffusion	扩散	drag due to flow around a body	绕流阻力
diffusion coefficient	扩散系数	drag, resistance	阻力
diffusion equation	扩散方程	drainage	排水
dilution	稀释度	dropdown curve	降水曲线
dimensional analysis	量纲分析	dropdown	跌水
dimensional quantities	有量纲量	dynamic pressure of flow	流体动压强
dimensionless quantities	无量纲量、量纲为一的量	dynamic similarity	动力相似性
		dynamic velocity (friction velocity)	动力速度（阻力速度）
dimensionless parameter	无量纲参数（更名为特征数）	dynamic viscosity	动力黏度

E

eddies	涡体	environmental fluid mechanics	环境流体力学
eddy	涡	equation of moment of momentum	动量矩方程
element flow	元流	equation of state	状态方程
element vortex	元涡	equilibrium	平衡
elevation head	位置水头	equipotential line	等势面
energy	能量	equipressure surface	等压面
energy dissipation	消能	equivalent diameter	当量直径
energy equation	能量方程	equivalent roughness	当量粗糙度
energy transfer	能量传递	Euler number	欧拉数
energy transport	能量运输	Euler's equilibrium equation	欧拉平衡方程
engineering atmospheric pressure	工程大气压强	Euler's motion equation	欧拉运动方程
engineering fluid mechanics	工程流体力学	experimental method	实验方法

F

falling slope	顺坡	fluid kinematic	流体运动学
finite volume method	有限体积法	fluid mechanics	流体力学
floating body	浮体	fluid particle	流体质点
flow coefficient	流量系数	force scale	力的比尺
flow cross-section	过水断面	free jet	自由射流
flow field	流场	free outflow	自由出流
flow in porous media, seepage flow	渗流	free surface	自由表面
flow net	流网	free surface flow	无压流
flow rate, flow discharge	流量	friction coefficient	摩擦系数
flow regime	流态	friction drag	摩擦阻力
fluctuating pressure	脉动压强	friction loss	摩擦损失
fluctuating velocity	脉动流速	Froude number	弗劳德数
fluctuation	脉动		

G

gas constant	气体常数	gradually-varying flow	渐变流
gasification	气化	gravitational water	重力水
geometric similarity	几何相似	groundwater	地下水

H

head lose	水头损失	hydraulic radius	水力半径
homogeneous fluid	均质流体	hydraulic slope	水力坡度
homogeneous soil	均质土壤	hydraulics	水力学
horizontal slope	平底坡	hydrodynamics	水动力学
hydraulic drop	水跌	hydrostatic pressure	液体静压
hydraulic jump	水跃	hydrostatics	水静力学

I

ideal fluid	理想流体	initial condition	初始条件
impulse	冲量	instability	不稳定性
incoming flow	来流	instantaneous velocity	瞬时速度
incompressible fluid	不可压缩流体	intensity of turbulence	湍流强度
indirect water hammer	间接水击	interface	分界面，界面
inertial force	惯性力	isotropy	各向同性

J

jet	射流	Joukowsky theorem	茹科夫斯基定理

K

kinematic similarity	运动相似	kinetic theory of gas	气体动理论
kinematic viscosity	运动黏度	kinetic-energy correction factor	动能修正系数
kinetic energy	动能		

L

laminar boundary layer	层流边界区	line source	线源
laminar flow	层流	linear deformation	线变形
length of jump	水跃长度	local acceleration	当地加速度
length scale	长度比尺	local atmospheric pressure	当地大气压强
line of seepage (depression line), seepage line	浸润曲线	long pipe	长管
		looping pipes	环状管网

M

Mach number	马赫数	mean velocity	平均速度
Manning formula	曼宁公式	mechanical energy	机械能
manometer	压强计	micromanometer	微压计
mass	质量	model	模型
mass force	质量力	model experiment	模型试验
mass transfer	质量传递	model scale	模型比尺
mathematical model	数学模型	modulus of elasticity	弹性模量

molecular diffusion	分子扩散	momentum	动量
molecular diffusion coefficient	分子扩散系数	momentum correction factor	动量修正系数
moment of force	力矩	momentum equation	动量方程
moment of inertia	转动惯量、惯性矩	momentum transfer	动量交换
		most efficient cross-section	水力最优断面
moment of momentum	动量距	multiple-well	井群

N

Navier-Stokes equation	纳维-斯托克斯方程	non-slip condition	无滑移条件
		non-uniform flow	非均匀流
negative pressure	负压	normal depth	正常水深
Newton number	牛顿数	normal stress	正应力
Newtonian fluid	牛顿流体	nonviscous fluid, inviscid fluid	无黏性流体
non-homogeneous fluid	非均质流体	nozzle flow	管嘴出流
non-Newtonian fluid	非牛顿流体	numerical calculation	数值计算
non-prismatic channel	非棱柱体渠道	numerical experiment	数值实验
nonscouring velocity	不冲流速	numerical simulation	数值模型
nonsilting velocity	不淤流速		

O

one dimensional flow	一维流动	orifice flow	孔口出流
open channel flow	明渠流	orifice meter	孔板流量计
orifice	孔口		

P

parallel flow	平行流	pollutant diffusion	污染物扩散
partially penetrating well	非完全井	pollutant source	污染源
path, path line	迹线	porosity	孔隙率
perfect gas	理想气体	porous medium	多孔介质
permissible velocity	允许流速	potential energy	势能,位能
pi theorem, Buckingham theorem; π theorem	π 定理	potential flow	势流
		potential force	有势力
piezometric head	测压管水头	power	功率
piezometric head line	测压管水头线	pressure	压强
pipe flow, tube flow	管流	pressure force	压力
pipe in parallel	并联管道	pressure drag	压差阻力
pipe in series	串联管道	pressure energy	压能
pipe networks	管网	pressure field	压强场
pipe with uniform discharge along the line	沿程均匀泄流管道	pressure flow	有压流
		pressure gage	压强表、压力表
Pitot tube	皮托管	pressure gradient	压强梯度
plane flow	平面流	pressure head	压强水头
plane jet	平面射流	pressure volume	压力体
plume	羽流	prismatic channel	棱柱体渠道
point of transition	过渡点	prototype	原型
point source	点源	pump	泵

Q

quantities of dimension one	量纲为一的量		

R

radius of influence	影响半径	relative velocity	相对速度
rapidly-varying flow	急变流	Reynolds number	雷诺数
rate of angular deformation	角变率	rising slope	逆坡
rate of linear deformation	线变率	rotational flow	有旋流、有涡流
region of square resistance law	阻力平方区	rough region	粗糙区
relative motion	相对运动	rough region of turbulent flow	湍流粗糙区
relative pressure	相对压强	roughness	粗糙度
relative roughness	相对粗糙度		

S

seepage pressure	渗透压强	stagnation point	驻点
seepage velocity	渗透速度	static head	静压水头
self-similar zone	自模区	static pressure	静压强
separation point	分离点	static pressure tube	静压管
settling velocity	沉降速度	steady flow	恒定流
shear flow	剪切流	steady plane potential flow	恒定平面势流
shear stress	切应力、剪应力	steep slope	陡坡、急坡
short pipe	短管	stratosphere	平流层
similar solution	相似性解	streak line	脉线、染色线
similarity conditions	相似条件	stream filament	流束
similarity criterion	相似准则	stream line	流线
similarity criterion number	相似判据、相似准数	stream tube	流管
		Strouhal number	施特鲁哈尔数
simulation	模拟	subcritical flow	缓流
singularity	奇点	submerged outflow	淹没出流
sink	汇	subsonic flow	亚声速流
siphon	虹吸管	supercritical flow	急流
smooth plate	光滑平板	supersonic flow	超声速流、超音速流动
smooth region of turbulent flow	湍流光滑区		
smooth wall	光滑壁面	surface force	表面力
specific energy	比能、断面单位能量	surface tension	表面张力
		system	系统
speed of sound	声速	systems of particles	质点系
stability	稳定性		

T

temperature gradient	温度梯度	three-dimensional flow	三维流
theorem of similarity (similar principle)	相似原理	time average value	时均值
		time scale	时间比尺
theory of dimensional homogeneity	量纲和谐原理	time-average method	时均法
		time-averaged concentration	时均浓度
thermodynamic temperature	热力学温度	total flow	总流

total head	总水头	turbulent boundary layer	湍流边界层
total head line	总水头线	turbulent core	湍流核心区
total pressure force	总压力	turbulent diffusion	湍流扩散
tracer	示踪物	turbulent diffusion coefficient	湍流扩散系数
transition layer	过渡层	turbulent flow	湍流流动
transition region of turbulent flow	湍流过渡区	turbulent fluctuation	湍流脉动
translation	平移	turbulent jet	湍流射流
transonic flow	跨声速流动	turbulent shear stress	湍流切应力
transport property	输运性质	two-dimensional flow	二维流
troposphere	对流层		

U

unconfined aquifer	无压含水层	unsteady flow	非恒定流、非定常流
uniform flow	均匀流		
unit	单位	U-tube	U 形管

V

vacuum pressure	真空压强	velocity scale	速度比尺
velocity circulation	速度环量	vena contract	收缩断面
velocity field	流速场	Venturi tube	文丘里管
velocity gradient	流速梯度	viscosity	黏度
velocity head	速度水头	viscous fluid	黏性流体
velocity of approach	行近流速	viscous sublayer	黏性底层
velocity potential	速度势		

W

wall roughness	壁面粗糙度	weight	重量
water hammer	水击、水锤	weir	堰
water level	水位	weir flow	堰流
wave height	波高	well	井
wave speed, wave velocity	波速	wetted perimeter	湿周
Weber number	韦伯数	work	功

References
参考文献

[1] Spurk J H. Fluid Mechanics [M]. 北京：世界图书出版公司北京公司，2001.
[2] Landau L D, Lifshitz E M. Fluid Mechanics [M]. 2nd ed. 北京：世界图书出版公司北京公司，1999.
[3] Douglas J F, Gasiorek J M, Swaffield J A. Fluid Mechanics [M]. 3rd ed. 北京：世界图书出版公司北京公司，2000.
[4] Streeter V L, Wylie E B, Bedford K W. Fluid Mechanics [M]. 9th ed. 北京：清华大学出版社，2003.
[5] Finnemore E J, Franzini J B. Fluid Mechanics with Engineering Applications [M]. 10th ed. 北京：清华大学出版社，2003.
[6] White F M. Fluid Mechanics [M]. 5th ed. 北京：清华大学出版社，2004.
[7] 郭仁东，冯劲梅，吴慧芳. 水力学 [M]. 北京：人民交通出版社，2006.
[8] 闻德荪，李兆年，黄正华. 工程流体力学：水力学 [M]. 2版. 北京：高等教育出版社，2004.
[9] 段文义，郭仁东，李亚峰. 水力学 [M]. 沈阳：东北大学出版社，2001.
[10] 王永岩，谢里阳，李庆领，等. 英汉双语流体力学电子教程 [M]. 北京：煤炭工业音像出版社，2004.
[11] 柯葵，李立明，李嵘. 水力学 [M]. 上海：同济大学出版社，2000.
[12] 吕文舫，柯葵. 水力学 [M]. 上海：同济大学出版社，1995.